209
Advances in Polymer Science

Editorial Board:
A. Abe · A.-C. Albertsson · R. Duncan · K. Dušek · W. H. de Jeu
J.-F. Joanny · H.-H. Kausch · S. Kobayashi · K.-S. Lee · L. Leibler
T. E. Long · I. Manners · M. Möller · O. Nuyken · E. M. Terentjev
B. Voit · G. Wegner · U. Wiesner

Advances in Polymer Science
Recently Published and Forthcoming Volumes

Functional Materials and Biomaterials
Vol. 209, 2007

Phase-Separated Interpenetrating Polymer Networks
Vol. 208, 2007

Hydrogen Bonded Polymers
Volume Editor: Binder, W.
Vol. 207, 2007

Oligomers · Polymer Composites · Molecular Imprinting
Vol. 206, 2007

Polysaccharides II
Volume Editor: Klemm, D.
Vol. 205, 2006

Neodymium Based Ziegler Catalysts – Fundamental Chemistry
Volume Editor: Nuyken, O.
Vol. 204, 2006

Polymers for Regenerative Medicine
Volume Editor: Werner, C.
Vol. 203, 2006

Peptide Hybrid Polymers
Volume Editors: Klok, H.-A., Schlaad, H.
Vol. 202, 2006

Supramolecular Polymers · Polymeric Betains · Oligomers
Vol. 201, 2006

Ordered Polymeric Nanostructures at Surfaces
Volume Editor: Vancso, G. J., Reiter, G.
Vol. 200, 2006

Emissive Materials · Nanomaterials
Vol. 199, 2006

Surface-Initiated Polymerization II
Volume Editor: Jordan, R.
Vol. 198, 2006

Surface-Initiated Polymerization I
Volume Editor: Jordan, R.
Vol. 197, 2006

Conformation-Dependent Design of Sequences in Copolymers II
Volume Editor: Khokhlov, A. R.
Vol. 196, 2006

Conformation-Dependent Design of Sequences in Copolymers I
Volume Editor: Khokhlov, A. R.
Vol. 195, 2006

Enzyme-Catalyzed Synthesis of Polymers
Volume Editors: Kobayashi, S., Ritter, H., Kaplan, D.
Vol. 194, 2006

Polymer Therapeutics II
Polymers as Drugs, Conjugates and Gene Delivery Systems
Volume Editors: Satchi-Fainaro, R., Duncan, R.
Vol. 193, 2006

Polymer Therapeutics I
Polymers as Drugs, Conjugates and Gene Delivery Systems
Volume Editors: Satchi-Fainaro, R., Duncan, R.
Vol. 192, 2006

Interphases and Mesophases in Polymer Crystallization III
Volume Editor: Allegra, G.
Vol. 191, 2005

Block Copolymers II
Volume Editor: Abetz, V.
Vol. 190, 2005

Functional Materials and Biomaterials

With contributions by
X. Dong Liu · A. R. Esker · M. Häußler · C. Kim · P. Lucas
M. Matsunaga · N. Nishi · J.-J. Robin · B. Z. Tang
D.-A. Wang · M. Yamada · H. Yu

 Springer

The series *Advances in Polymer Science* presents critical reviews of the present and future trends in polymer and biopolymer science including chemistry, physical chemistry, physics and material science. It is adressed to all scientists at universities and in industry who wish to keep abreast of advances in the topics covered.

As a rule, contributions are specially commissioned. The editors and publishers will, however, always be pleased to receive suggestions and supplementary information. Papers are accepted for *Advances in Polymer Science* in English.

In references *Advances in Polymer Science* is abbreviated *Adv Polym Sci* and is cited as a journal.

Springer WWW home page: springer.com
Visit the APS content at springerlink.com

Library of Congress Control Number: 2007926112

ISSN 0065-3195
ISBN 978-3-540-71508-5 Springer Berlin Heidelberg New York
DOI 10.1007/978-3-540-71509-2

This work is subject to copyright. All rights are reserved, whether the whole or part of the material is concerned, specifically the rights of translation, reprinting, reuse of illustrations, recitation, broadcasting, reproduction on microfilm or in any other way, and storage in data banks. Duplication of this publication or parts thereof is permitted only under the provisions of the German Copyright Law of September 9, 1965, in its current version, and permission for use must always be obtained from Springer. Violations are liable for prosecution under the German Copyright Law.

Springer is a part of Springer Science+Business Media

springer.com

© Springer-Verlag Berlin Heidelberg 2007

The use of registered names, trademarks, etc. in this publication does not imply, even in the absence of a specific statement, that such names are exempt from the relevant protective laws and regulations and therefore free for general use.

Cover design: WMXDesign GmbH, Heidelberg
Typesetting and Production: LE-TEX Jelonek, Schmidt & Vöckler GbR, Leipzig

Printed on acid-free paper   02/3100 YL – 5 4 3 2 1 0

Editorial Board

Prof. Akihiro Abe
Department of Industrial Chemistry
Tokyo Institute of Polytechnics
1583 Iiyama, Atsugi-shi 243-02, Japan
aabe@chem.t-kougei.ac.jp

Prof. A.-C. Albertsson
Department of Polymer Technology
The Royal Institute of Technology
10044 Stockholm, Sweden
aila@polymer.kth.se

Prof. Ruth Duncan
Welsh School of Pharmacy
Cardiff University
Redwood Building
King Edward VII Avenue
Cardiff CF 10 3XF, UK
DuncanR@cf.ac.uk

Prof. Karel Dušek
Institute of Macromolecular Chemistry,
Czech
Academy of Sciences of the Czech Republic
Heyrovský Sq. 2
16206 Prague 6, Czech Republic
dusek@imc.cas.cz

Prof. W. H. de Jeu
FOM-Institute AMOLF
Kruislaan 407
1098 SJ Amsterdam, The Netherlands
dejeu@amolf.nl
and Dutch Polymer Institute
Eindhoven University of Technology
PO Box 513
5600 MB Eindhoven, The Netherlands

Prof. Jean-François Joanny
Physicochimie Curie
Institut Curie section recherche
26 rue d'Ulm
75248 Paris cedex 05, France
jean-francois.joanny@curie.fr

Prof. Hans-Henning Kausch
Ecole Polytechnique Fédérale de Lausanne
Science de Base
Station 6
1015 Lausanne, Switzerland
kausch.cully@bluewin.ch

Prof. Shiro Kobayashi
R & D Center for Bio-based Materials
Kyoto Institute of Technology
Matsugasaki, Sakyo-ku
Kyoto 606-8585, Japan
kobayash@kit.ac.jp

Prof. Kwang-Sup Lee
Department of Polymer Science &
Engineering
Hannam University
133 Ojung-Dong
Daejeon 306-791, Korea
kslee@hannam.ac.kr

Prof. L. Leibler
Matière Molle et Chimie
Ecole Supérieure de Physique
et Chimie Industrielles (ESPCI)
10 rue Vauquelin
75231 Paris Cedex 05, France
ludwik.leibler@espci.fr

Prof. Timothy E. Long
Department of Chemistry
and Research Institute
Virginia Tech
2110 Hahn Hall (0344)
Blacksburg, VA 24061, USA
telong@vt.edu

Prof. Ian Manners
School of Chemistry
University of Bristol
Cantock's Close
BS8 1TS Bristol, UK
ian.manners@bristol.ac.uk

Prof. Martin Möller
Deutsches Wollforschungsinstitut
an der RWTH Aachen e.V.
Pauwelsstraße 8
52056 Aachen, Germany
moeller@dwi.rwth-aachen.de

Prof. Oskar Nuyken
Lehrstuhl für Makromolekulare Stoffe
TU München
Lichtenbergstr. 4
85747 Garching, Germany
oskar.nuyken@ch.tum.de

Prof. E. M. Terentjev
Cavendish Laboratory
Madingley Road
Cambridge CB 3 OHE, UK
emt1000@cam.ac.uk

Prof. Brigitte Voit
Institut für Polymerforschung Dresden
Hohe Straße 6
01069 Dresden, Germany
voit@ipfdd.de

Prof. Gerhard Wegner
Max-Planck-Institut
für Polymerforschung
Ackermannweg 10
Postfach 3148
55128 Mainz, Germany
wegner@mpip-mainz.mpg.de

Prof. Ulrich Wiesner
Materials Science & Engineering
Cornell University
329 Bard Hall
Ithaca, NY 14853, USA
ubw1@cornell.edu

Advances in Polymer Science
Also Available Electronically

For all customers who have a standing order to Advances in Polymer Science, we offer the electronic version via SpringerLink free of charge. Please contact your librarian who can receive a password or free access to the full articles by registering at:

springerlink.com

If you do not have a subscription, you can still view the tables of contents of the volumes and the abstract of each article by going to the SpringerLink Homepage, clicking on "Browse by Online Libraries", then "Chemical Sciences", and finally choose Advances in Polymer Science.

You will find information about the

- Editorial Board
- Aims and Scope
- Instructions for Authors
- Sample Contribution

at springer.com using the search function.

Contents

Functional Hyperbranched Macromolecules Constructed from Acetylenic Triple-Bond Building Blocks
M. Häußler · B. Z. Tang 1

Polymer Monolayer Dynamics
A. R. Esker · C. Kim · H. Yu 59

Silicone-Based Polymer Blends: An Overview of the Materials and Processes
P. Lucas · J.-J. Robin 111

Functional Materials Derived from DNA
X. Dong Liu · M. Yamada · M. Matsunaga · N. Nishi 149

Engineering Blood-Contact Biomaterials by "H-Bond Grafting" Surface Modification
D.-A. Wang 179

Author Index Volumes 201–209 229

Subject Index 233

Functional Hyperbranched Macromolecules Constructed from Acetylenic Triple-Bond Building Blocks

Matthias Häußler · Ben Zhong Tang (✉)

Department of Chemistry, The Hong Kong University of Science & Technology, Clear Water Bay, Kowloon, Hong Kong,
and Department of Polymer Science and Engineering, Zhejiang University, 310027 Hangzhou, China
tangbenz@ust.hk

1	Introduction	4
2	Synthesis	4
2.1	Polycoupling	5
2.1.1	Palladium-Catalyzed Polycoupling	5
2.1.2	Mechanistic Considerations	8
2.1.3	Copper-Catalyzed Polycoupling	9
2.2	Polyhydrosilylation	12
2.2.1	Platinum-Catalyzed Polyhydrosilylation	13
2.2.2	Rhodium- and Palladium-Catalyzed Polyhydrosilylation	13
2.3	Polycycloaddition	14
2.3.1	Diels–Alder Polycycloaddition	14
2.3.2	1,3-Dipolar Polycycloaddition (Click Polymerization)	15
2.4	Polycyclotrimerization	19
2.4.1	Hyperbranched Poly(alkylenephenylene)s	20
2.4.2	Hyperbranched Polyarylenes	26
2.4.3	Hyperbranched Poly(aroylarylene)s	33
3	Properties	37
3.1	Thermal and Optical Properties	37
3.2	Patterning and Assembling Behaviors	43
3.3	Photonic, Magnetic and Catalytic Properties	48
4	Concluding Remarks	53
References		54

Abstract This review article summarizes the synthetic efforts in constructing functional hyperbranched macromolecules from acetylenic triple-bond building blocks. Polymerization reactions including polycoupling, polyaddition and polycyclotrimerization have been developed for the synthesis of new hyperbranched polymers such as polyynes, polyenes, polyarylenes and polytriazoles with novel topological structures and electronic conjugations. Polymers with high molecular weights (up to $> 1 \times 10^6$) have been obtained in high yields (up to 100%). Whilst their linear counterparts are often intractable, the hyperbranched conjugated polymers are completely soluble in common organic solvents and are hence readily processable by macroscopic techniques. The hyperbranched polymers exhibit an array of functional properties including strong light emission, stable optical nonlinearity and high photorefractivity. The polymers can generate fluorescent images, assemble into supramolecular patterns, and form well-aligned nanotubes. The polyynes can be post-functionalized through metal complexation, whose refractive indexes can be manipulated to a great extent by photoirradiation. The hyperbranched polymer complexes can serve as precursors to soft ferromagnetic ceramics and as catalysts for carbon nanotube fabrications.

Keywords Functional materials · Hyperbranched polymers · Polyaddition · Polycoupling · Polycyclotrimerization

Abbreviations
AAO	anodic aluminum oxide
AFM	atomic force microscopy
Ar	aromatic ring
β	molecular first hyperpolarizability
Bu	butyl group
c	concentration
C	core molecule
CNT	carbon nanotube
Cp*	pentamethylcyclopentadienyl ligand
CVD	chemical vapor deposition
D	dendritic unit
$D_{0.5}$	dose needed for an F_g value of 0.5 (or 50%)
d_{33}	macroscopic SHG coefficient
DB	degree of branching
DCM	dichloromethane
D_e	exposure dose
DMF	N,N-dimethylformamide
DSC	differential scanning calorimetry
EO	electro-optical
F_g	gel fraction
F_{OL}	optical limiting threshold fluence
FTIR	Fourier-transfer infrared spectroscopy
$F_{t,m}/F_{i,m}$	optical signal suppression ratio
GPC	gel permeation chromatography
H	strength of magnetic field
hb-PA	hyperbranched polyarylene
hb-PAA	hyperbranched poly(aroylarylene)
hb-PAE	hyperbranched poly(aryleneethynylene)
hb-PAP	hyperbranched poly(alkylenephenylene)
hb-PP	hyperbranched polyphenylene
hb-PPE	hyperbranched poly(phenyleneethynylene)
hb-PTA	hyperbranched polytriazole
hb-PY	hyperbranched polyyne
H_c	coercivity
L	ligand
L	linear unit
l_f	film thickness
λ_{ab}	absorption maximum

λ_{em}	emission maximum
λ_{ex}	excitation wavelength
M	magnetization
M	monomer
Me	methyl group
M_n	number-average molecular weight
M_s	saturation magnetization
M_w	weight-average molecular weight
MWD	molecular weight distribution
NLO	Nonlinear optical
NMR	nuclear magnetic resonance spectroscopy
P	polymer
PB	polymer branch
PC	polycarbonate
PDI	polydispersity index (M_w/M_n)
PEO_{600}	poly(ethylene oxide) with a molecular weight of 600
pH	solution acidity
PH	poly(1-hexyne)
Φ_F	fluorescence quantum yield
pK_a	acidity constant
PL	photoluminescence
PMMA	poly(methyl methacrylate)
PPA	poly(phenylacetylene)
PPP	poly(p-phenylene)
PS	polystyrene
RI	refractive index (n)
S	solid support
SEC	size-exclusion chromatography
SEM	scanning electron microscope
SHG	second-harmonic generation
t	exposure time
T	terminal unit
T	terminator
T_d	thermal degradation temperature
TEM	transmission electron microscope
T_g	glass transition temperature
TGA	thermogravimetric analysis
THF	tetrahydrofuran
T_L	linear transmittance
TMEDA	N,N,N',N'-tetramethylethylenediamine
TPA	triphenylamine
UV	ultraviolet
W_r	weight residue after pyrolysis
X	halogen
XPS	X-ray photoelectron spectroscopy
XRD	X-ray diffraction

1
Introduction

Hyperbranched polymers have emerged as a new class of macromolecules that show architectural beauty and multifaceted functionality of dendrimers while enjoying the ease of being prepared by simple, single-step reaction procedures. A number of strategies have been developed for the synthesis of hyperbranched polymers. The commonly adopted approach is self-condensation polymerization of AB_x-type monomers with $x \geq 2$ where A and B are mutually reactive functional groups, dating back to the theoretical work of Flory in the early 1950s [1]. Because of the limited commercial availability and difficult synthetic access to multifunctional monomers bearing multiple, mutually reactive groups, alternative approaches such as copolymerizations of A_2 monomers with B_x comonomers ($x \geq 3$) have been developed [2–7]. Other polymerization reactions including self-condensing vinyl polymerizations initiated by cationic [8] and radical catalysts [9, 10] and ring-opening multibranching polymerizations [11–15] have been explored, mainly for the synthesis of non-conjugated hyperbranched polymers [16–18].

Hyperbranched macromolecules have been constructed from various functional groups, among which, carbon-carbon triple-bond functionality uniquely stands out because it offers ready access to hyperbranched conjugative macromolecules. Being unsaturated, it accommodates various addition reactions. In comparison to vinyl and alkyl protons, the acetylenic proton is most acidic ($pK_a = 26$; cf., $pK_a = 45$ for ethylene and $pK_a = 62$ for ethane), thus enabling facile substitution and coupling reactions. In this account, we give a brief overview of the synthesis of new hyperbranched polymers from acetylenic monomers, with an emphasis on our recent effort devoted to the development of new synthetic routes to conjugated hyperbranched polymers. Examples of their advanced functional properties are also presented.

2
Synthesis

Using acetylenic monomers carrying single and multiple triple bonds, a variety of new hyperbranched polymers with unique structures and novel properties have been synthesized by the alkyne polymerization reactions such as polycoupling, polyhydrosilylation, polycycloaddition, and polycyclotrimerization initiated by transition-metal catalysts (e.g., tantalum halides) and non-metallic species (e.g., secondary amines). Through molecular structural design and reaction condition optimization, hyperbranched polymers with high molecular weights and ready macroscopic processability have been obtained in high yields.

2.1
Polycoupling

Homo- and cross-couplings of alkyne promoted by metallic catalysts are versatile reactions for carbon-carbon bond formation and have been utilized to synthesize functional hyperbranched polymers.

2.1.1
Palladium-Catalyzed Polycoupling

Repetitive coupling of acetylenes with aryl halides is an effective way to directly build hyperbranched architecture in a stepwise manner. This type of polycoupling is often catalyzed by palladium complexes in the presence of amines and has been widely used for the preparation of well-defined oligomers, linear polymers, and perfectly branched dendrimers [19, 20].

Employing this reaction protocol, hb-PAEs, particularly hb-PPEs, have been prepared. Diiodophenylacetylene (1) represents a typical AB_2-type building block in this approach (Scheme 1). Palladium-catalyzed polycoupling of 1 furnished an hb-PPE with a bimodal SEC elution profile, charac-

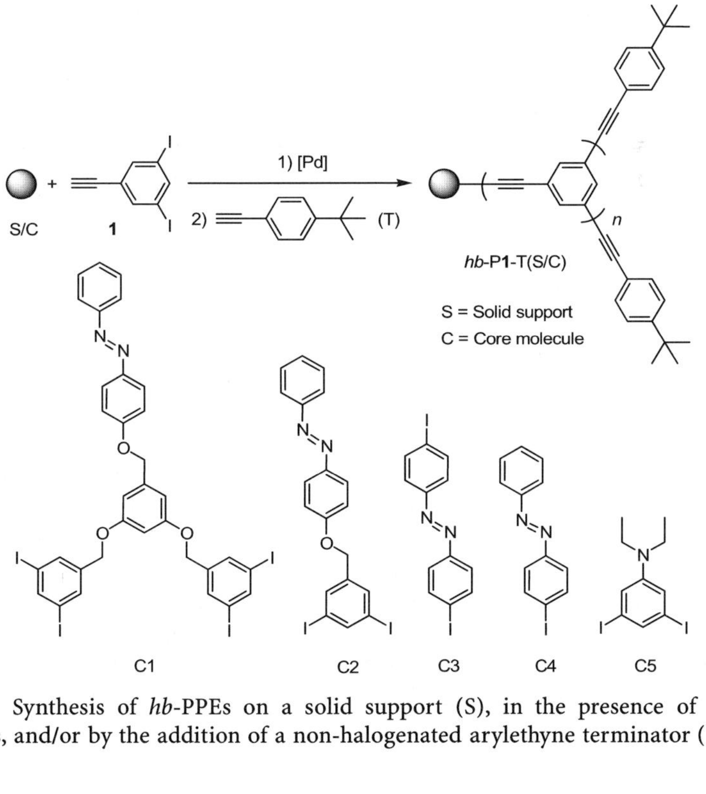

Scheme 1 Synthesis of hb-PPEs on a solid support (S), in the presence of core (C) molecules, and/or by the addition of a non-halogenated arylethyne terminator (T)

terized by a relatively narrow peak at the very high molecular weight region along with a long tail extending to the low molecular weight region [21]. In contrast, the *hb*-PPE obtained from the solid-supported polymerization under similar conditions possessed monomodal, narrow MWD or small PDI (M_w/M_n = 1.3). The propagation reaction could be controlled by changing the nature of the support (e.g. cross-linking density and loading) to a certain extent.

Slow addition of **1** to the solutions of multifunctional core molecules (C1–C5) resulted in higher molecular weight *hb*-PPEs with again small PDI values [22]. Control over the molecular weight was achieved by varying the monomer/core ratio. The PDI values of the *hb*-PPEs were found to decrease with increasing the degree of polymerization and with increasing the number of functional groups of the core molecule.

The Pd-polycoupling protocol works for other monomers of similar structures. For example, an AB_2-type monomer containing a 1,3,5-triazine moiety (**2**) was successfully converted into an *hb*-PAE (Scheme 2), although the resulting *hb*-P2 was structurally irregular due to the formation of diacetylene branches in the polycoupling reaction [23]. Another example is the synthesis of a luminescent hyperbranched polymer from the polycoupling of β,β-dibromo-4-ethynylstyrene **3** (Scheme 3) [24]. Although the resulting *hb*-P3 was only partially soluble, the structure could be elucidated by spectroscopic methods with the aid of a model compound. Theoretical calculations revealed a partially disrupted conjugation structure due to the twist of the benzene rings.

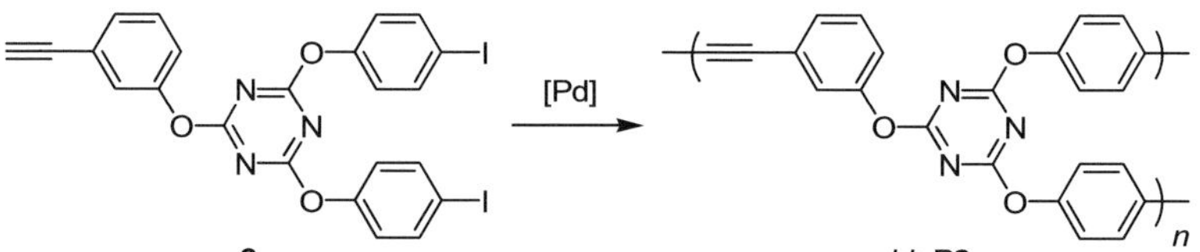

Scheme 2 Synthesis of an *hb*-PAE containing a triazine core

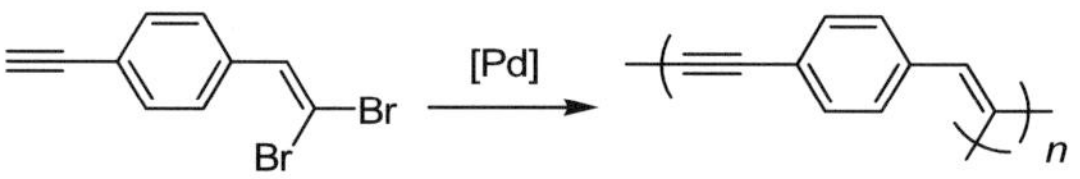

Scheme 3 Synthesis of hyperbranched polymer from β,β-dibromo-4-ethynylstyrene

Putting both halide and acetylene functionalities into a single AB_2 monomer is not a trivial job due to the involved synthetic difficulty and problems such as the formation of isomeric mixtures and undesired self-oligomerization. Purification of the desired monomers is often troublesome and ends with low isolation yields. Separating halide and acetylene groups into two different molecules allows easier synthetic access and offers a greater variety of monomer choices but meanwhile invites the risk of cross-linking reactions. Control of the polycoupling conditions is thus a necessity if one intends to synthesize soluble hyperbranched polymers with desired structures and properties.

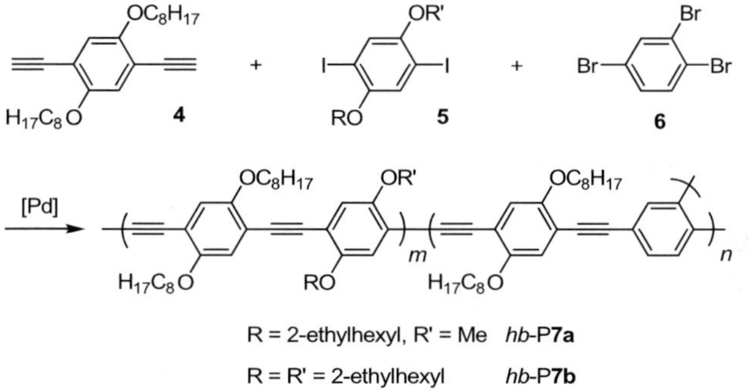

Scheme 4 Formation of cross-linked particles in the polymerization of $A_2 + B_2 + B_3$ monomers

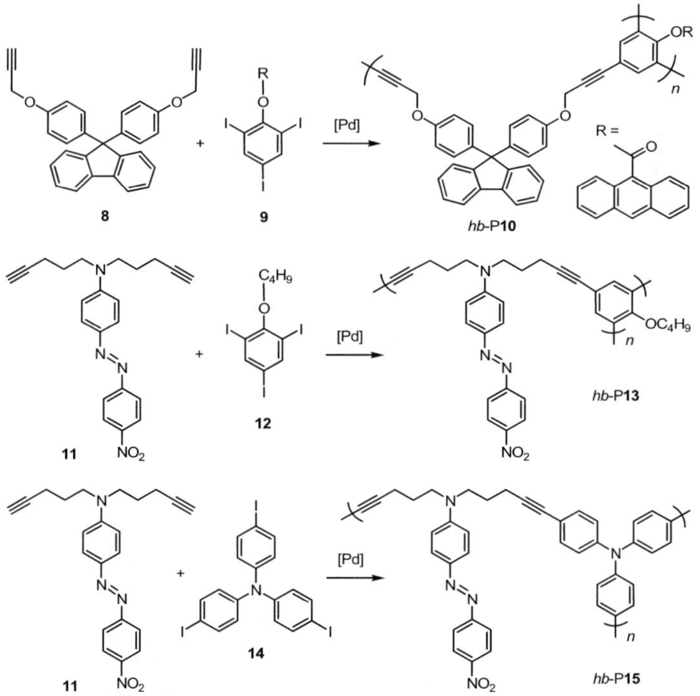

Scheme 5 $A_2 + B_3$ approach toward soluble, functional hb-PAEs

Cross-linking reactions readily occurred in the polymerizations of A_2 + B_2 + B_3 monomers in aqueous emulsions (Scheme 4) [25, 26]. The size of the cross-linked particles was dependant on the polymerization conditions. Micrometer-sized particles were formed when the reaction was carried out in a water–toluene–diisopropylamine mixture, whereas smaller nanoparticles were formed when the reaction was conducted in an ultrasonic bath.

We tried to optimize the polycoupling conditions by varying such parameters as polymerization time, monomer concentration and monomer addition mode, in an effort to control the polymer formation and to render the polymers soluble and processable. The optimization worked well and our A_2 + B_3 approach offered ready access to a soluble *hb*-PAE containing luminescent anthracene and fluorene chromophores (Scheme 5) [27]. Similarly, soluble azo-functionalized polymers *hb*-P13 and *hb*-P15 were obtained from the palladium-catalyzed polycoupling of triiodoarenes (12 and 14) with a diethynylazobenzene (11) [28].

2.1.2
Mechanistic Considerations

The overall reactions for the formation of *hb*-PAEs via Pd-catalyzed coupling are depicted in Scheme 6. As discussed above, two synthetic strategies are currently employed. The first one utilizes AB_2-type monomers, building up hyperbranched architecture through repetitive coupling of the triple bond with aryl halide. If no side reactions occur, this protocol allows only one single internal cyclization of an aryl halide with the focal acetylene unit, thus yielding *hb*-PAEs without any cross-link points. On the other hand, this internal ring closure leads to the formation of various polymeric species with different propagation possibilities, which are the cause for an increased MWD at higher conversions. Terminating the focal unit by core molecules can nicely overcome this problem (cf., Scheme 1) or even reverse the trend: with an increase in the conversion, the MWD becomes narrower [21, 22].

The second method separates the functional groups into two monomers, which facilitates synthetic work and offers greater choices to monomeric structure. In the first step, A_2 and B_3 monomers couple together to form an AB_2-type dimer that continues to react to form the hyperbranched architecture (Scheme 6). This is the case, only if the molar ratio of A_2 to B_3 is 1 : 1 and the initiation is considerably faster than the propagation [29]. It becomes immediately clear that the resultant structure is highly dependant on the type of monomers and the polymerization conditions. For the latter, it has been found that the mode of monomer addition plays a crucial role. Whereas the addition of a B_3 monomer into a solution of A_2 yields insoluble polymer gel, the opposite addition mode furnishes hyperbranched polymers with excellent solubility [30].

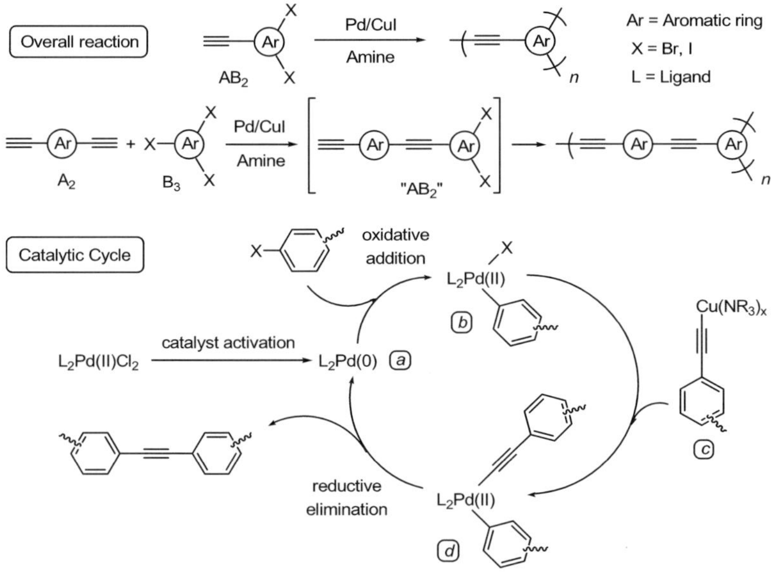

Functional Hyperbranched Macromolecules

Scheme 6 Mechanisms for Pd-catalyzed polycoupling reactions

The catalytic cycle for the Pd-catalyzed polycoupling is shown in the lower panel of Scheme 6. In many cases, Pd(II) complexes such as the commercially available $(Ph_3P)_2PdCl_2$ are used as the catalytic source of Pd. In the first step, the Pd(II) is activated by cuprated alkynes (such as c) under reductive butadiyne elimination, creating the active catalytic species a. Oxidative addition of the aromatic halide produces intermediate b, which after transmetalation with c gives diorganopalladium species d. This species undergoes reductive elimination to finish the coupling cycle and re-forms the active catalyst a [31].

2.1.3
Copper-Catalyzed Polycoupling

Acetylene triple bonds react with not only aryl halides but also transition-metal complexes. An example is shown in Scheme 7, where 1,3,5-triethynylbenzene is functionalized by two equivalents of $Pt(PBu_3)Cl_2$ to form an organometallic AB_2-type monomer (**16**). Self-polycoupling of **16** initiated by a copper(I) catalyst yielded hyperbranched polymer hb-P**16**, which was soluble in common organic solvents and could be fully characterized by spectroscopic and chromatographic methods including SEC [32]. Cross-polycoupling between a Pt complex and an organic triyne via $A_2 + B_3$ protocol resulted in the formation of a product that was insoluble in common solvents. Only with the addition of a very large excess (e.g., 50 fold) of 1,4-diethynylbenzene, could the involved cross-linking reactions be depressed [33].

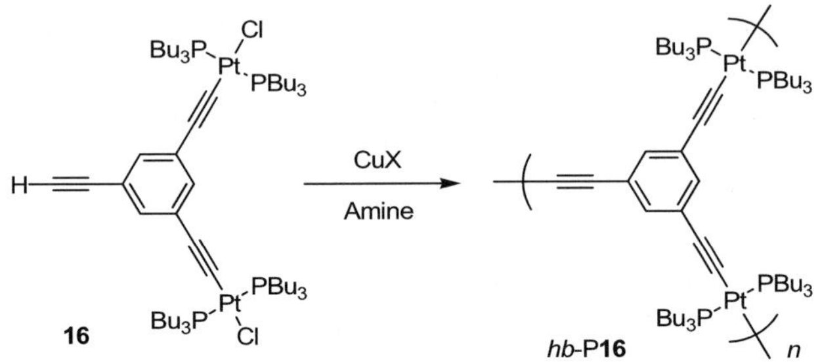

Scheme 7 Synthesis of hyperbranched organometallic polymer from AB$_2$-type monomer

Using aromatic triynes Ar(C≡CH)$_3$ as monomeric building blocks, *hb*-PYs can be readily constructed (Scheme 8) [34]. The fast polycoupling reaction could be controlled by terminating the propagating species before the

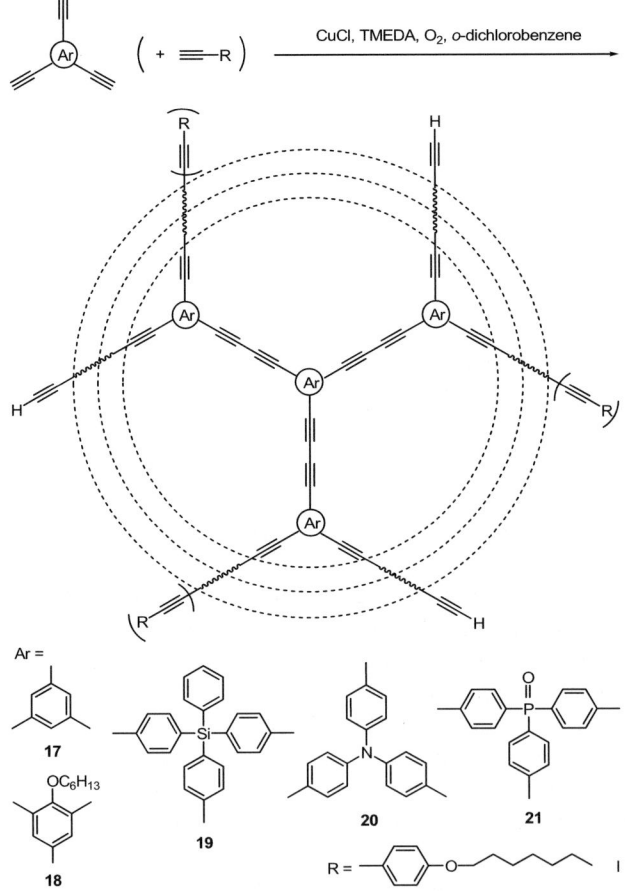

Scheme 8 Synthesis of *hb*-PYs through (co)polycoupling of triynes (with monoynes)

gel point by simply pouring the reaction mixture into acidified methanol. Soluble, high molecular weight hb-PYs of triynes like **17** and **19**, whose Cu-catalyzed homopolycoupling proceeded very fast, could be obtained by their copolycoupling with monoynes such as I.

In the polycoupling reactions, the formation of the diyne units proceeded via a Glaser–Hay oxidative coupling route [35–38]. Despite its wide applications in the preparation of small molecules and linear polymers containing diyne moieties, its mechanism remains unclear [38–40]. It has been proposed that a dimeric copper acetylide complex is involved, whose collapse leads to the formation of the diyne product (Scheme 9).

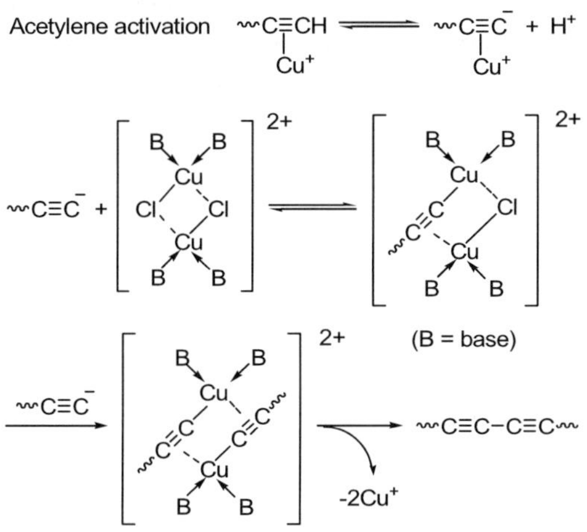

Scheme 9 Acetylene activation via π-complexation with copper(I) and proposed mechanism for the formation of diacetylene bond via cuprated alkyne dimers

Because of the one-step polymerization procedure, hyperbranched polymers often contain not only D and T but also L repeating units. This can be expressed by DB, which is an important structural parameter of hyperbranched polymers. DB is estimated as the sum of the D and T units divided by the sum of all the three structural units, that is, D, T and L [41]. By definition, a linear polymer has no dendritic units and its DB is zero, while a perfect dendrimer has no linear units and its DB is thus unity. Frey has pointed out that DB statistically approaches 0.5 in the case of polymerization of AB_2 monomers, provided that all the functional groups possess the same reactivity [42]. The structures of the hb-PYs could be analyzed by spectroscopic methods such as NMR and FTIR. The DB value of the phosphorous-containing polymer hb-**P21**, for example, was estimated to be 53% from its ^{31}P NMR chemical shifts (Chart 1).

The spectroscopic analyses revealed that both the homo- and copolyynes contained terminal triple bonds, which offers a nice opportunity to decorate

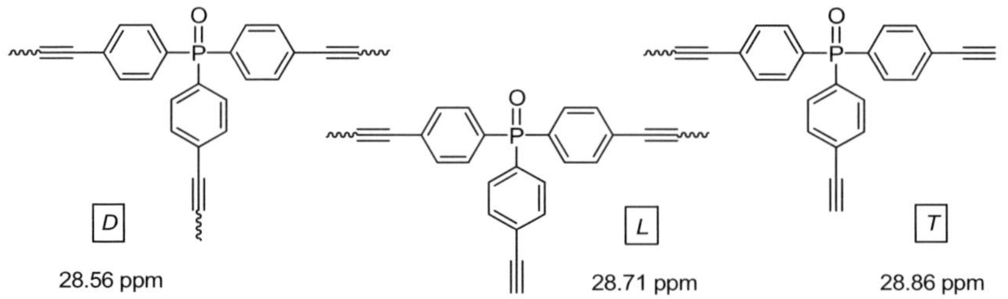

Chart 1 Structures of dendrtic (*D*), linear (*L*) and terminal (*T*) units in *hb*-P21 and their ^{31}P NMR chemical shifts

the peripheries of the polymers by end-capping reactions. This is demonstrated by the coupling of *hb*-P20 with aryl iodides II and III (Scheme 10). The polymer capped by phenyl iodide (*hb*-P20-II) became partially soluble after purification, possibly due to the π–π stacking-induced supramolecular aggregation during the precipitation and drying processes. Product *hb*-P20-III was soluble, thanks to the long dodecyloxy group of the end-capping agent. No signal of terminal acetylene resonance was observed in its ^{1}H NMR spectrum, unambiguously confirming the completion of the end-capping reaction.

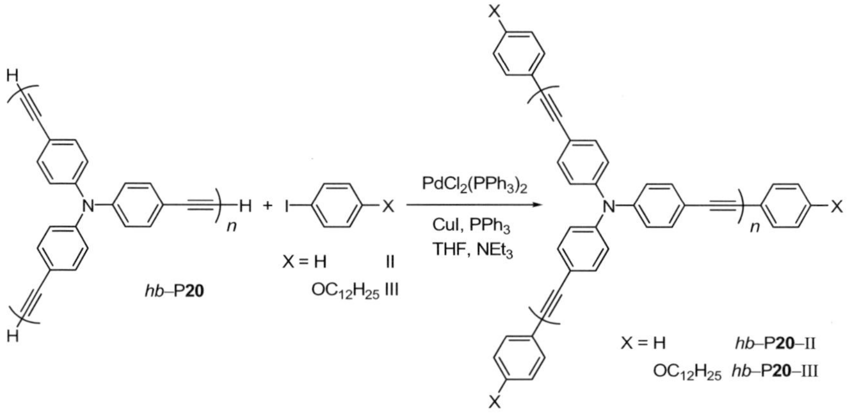

Scheme 10 End-capping reactions of *hb*-P20

2.2
Polyhydrosilylation

Similar to olefins, acetylenes undergo facile addition reactions. Transition metal-catalyzed hydrosilylation is such a reaction that has been utilized for the synthesis of silicon-containing hyperbranched polymers.

2.2.1
Platinum-Catalyzed Polyhydrosilylation

A hyperbranched polycarbosilane (*hb*-P22) was prepared by platinum-catalyzed polyhydrosilylation of methyldiethynylsilane 22 (Scheme 11) [43]. The tacky, highly soluble and stable polymer underwent thermo- and photo-induced cross-linking reactions through the peripheral ethynyl groups.

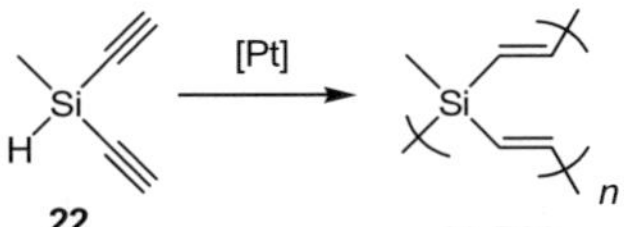

Scheme 11 Synthesis of hyperbranched polycarbosilane via polyhydrosilylation

2.2.2
Rhodium- and Palladium-Catalyzed Polyhydrosilylation

Regio- and stereoregular σ-π-conjugated hyperbranched polymers were prepared from rhodium-catalyzed polyhydrosilylation of AB$_2$-type silane monomers 23 and 24 (Scheme 12) [44, 45]. Polymer *hb*-P23 contained 95% trans vinylene units and lost merely 9% of its weight when heated under nitrogen to a temperature as high as 900 °C. It showed a UV absorption peak at ~ 275 nm due to the π–π^* transition and a weak absorption at ~ 330 nm due to the π-to-σ charge transfer, which was hardly seen in its linear polymer congeners.

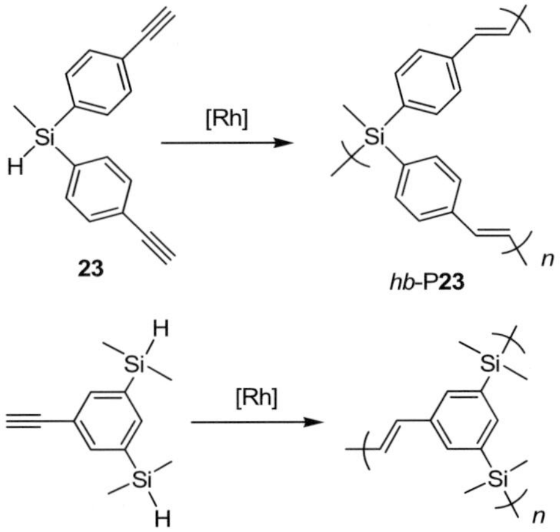

Scheme 12 Synthesis of σ-π conjugated hyperbranched polymers through rhodium-catalyzed polyhydrosilylation

The unreacted peripheral silane groups of *hb*-P24 could either be terminated by reacting with an excess amount of phenylacetylene or cross-linked by hydrolysis and aerial oxidation. Whereas the first reaction gave a soluble hyperbranched polymer, the latter furnished a transparent gel via siloxane linkages without any obvious defects such as phase separation and decomposition. The gel exhibited a strong resistance against various organic solvents, showed a weak absorption at above 300 nm due to the involved charge transfer, and emitted an intense blue light.

In a similar approach, palladium-catalyzed polyhydrosilylation afforded soluble hyperbranched poly(silylenedivinylene)s through an $A_2 + B_2 + B_3$ protocol with 1,3,5-triethynylbenzene and B,B',B''-triethynyl-N,N',N''-trimethylborazine as the branching units [46]. The resultant polymers showed higher thermal stabilities and higher char yields than their linear analogs. The hyperbranched polymers obtained from 1,3,5-triethynylbenzene showed intensified broad UV absorptions and enhanced light emissions.

2.3
Polycycloaddition

Polycycloaddition reactions of acetylenic monomers have been used to synthesize hyperbranched polymers comprised of pure aromatic rings and heterocyclic units.

2.3.1
Diels–Alder Polycycloaddition

Müllen and coworkers utilized the Diels–Alder reaction to prepare *hb*-PPs (*hb*-P25 to *hb*-P27) with M_w up to 107 000 from ethynyl-, propynyl-, and phenylethynyl-substituted tetraphenylcyclopentadienones (Scheme 13) [47]. Despite their high molecular weights and rigid phenyl rings, all the polymers showed good solubility in common aromatic solvents such as toluene and benzene.

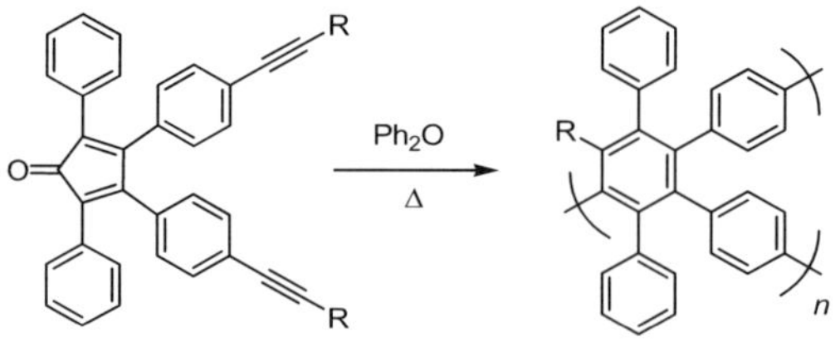

Scheme 13 Synthesis of *hb*-PPs via Diels–Alder polycycloaddition

2.3.2
1,3-Dipolar Polycycloaddition (Click Polymerization)

The formation of triazole rings by the reactions of azides and acetylenes was first described by Huisgen and coworkers [48, 49] and has recently been promoted as "click" chemistry by Sharpless et al. [50, 51]. This versatile [3 + 2] dipolar cycloaddition proved to be useful for the synthesis of hyperbranched polytriazoles via 1,3-dipolar polycycloaddition of AB_2-type monomers **28** and **29** (Scheme 14) [52]. The monomers exhibited very high reactivity: the

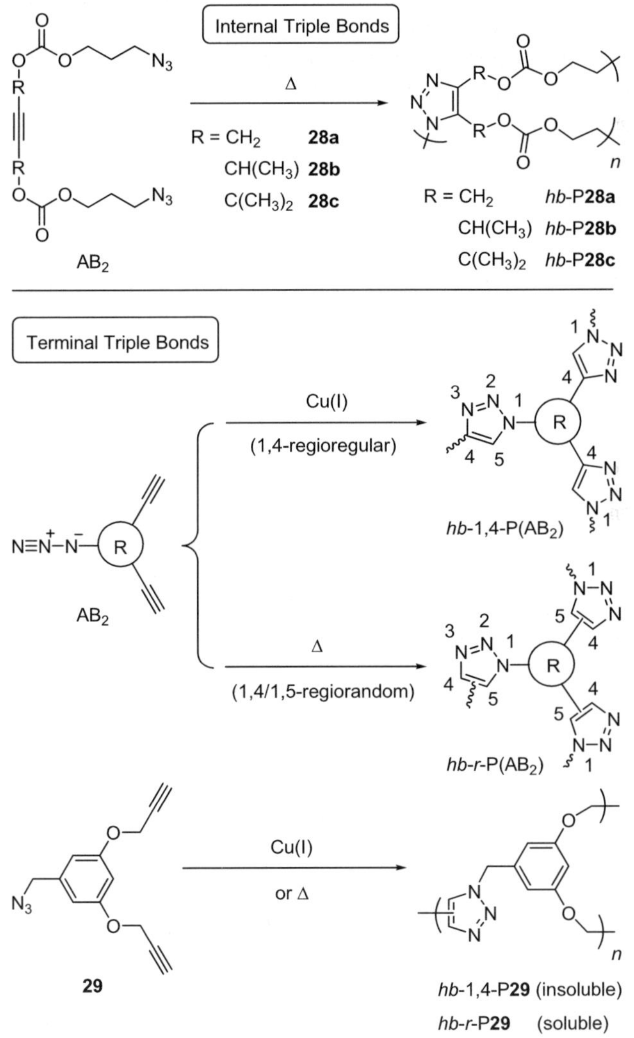

Scheme 14 Synthesis of *hb*-PTAs by "click" polymerization from AB_2-type monomers

low temperature auto-polymerization transformed them into soluble, high molecular weight *hb*-PTAs with an estimated DB value of 50%. Copper(I)-catalyzed polycycloaddition of **29** furnished a 1,4-stereoregular *hb*-1,4-P**29**, however, with intractability.

This study demonstrated the potential of the [3 + 2] polycycloaddition reaction involving azides and alkynes for the preparation of new hyperbranched polymers and meanwhile showed the necessity of controlling the polymerization conditions because of the high reactivity of the azide and alkyne functionalities [53]. One possibility to gain control and to inhibit autopolymerization of the monomers is to put the functional groups in two separate monomers. This would make it easier to synthesize and purify the monomers and make it possible to extend the shelf-life of the monomers. Thermally induced 1,3-dipolar polycycloaddition of diazides and triynes furnished soluble hyperbranched polytriazoles (Scheme 15 and Table 1, entries 1 and 2) [54]. Similar to the AB$_2$-type monomers, attempts to prepare 1,4-regioregular polytriazoles through copper(I)-mediated cycloaddi-

Scheme 15 *hb*-PTAs via A$_2$ + B$_3$ 1,3-dipolar cycloaddition

Table 1 Synthesis of hyperbranched polytriazoles

Entry	Monomers	Time (h)	Polymer	Yield (%)	M_w[a]	PDI[a]
Thermally Initiated Polymerization[b]						
1	30(4) + 20	72	hb-r-P30(4)-20	64.0	5500	2.0
2	30(6) + 20	72	hb-r-P30(6)-20	75.7	11400	2.7
Transition Metal-Catalyzed Polymerization[b]						
3	30(4) + 20	0.33	hb-1,5-P30(4)-20	62.5[c]	5350	2.4
4	30(6) + 20	0.50	hb-1,5-P30(6)-20	74.9	9370	2.7

[a] Estimated in THF by GPC on the basis of a polystyrene calibration
[b] $[20]_0/[30(m)]_0 = 2:3$, $[20]_0 = 0.12$ M; under nitrogen in 1,4-dioxane at 101 °C (entries 1 and 2) or in THF at 60 °C using Cp*Ru(PPh$_3$)$_2$Cl as catalyst (entries 3 and 4)
[c] Soluble fraction (total yield: 84.5%)

tion resulted only in insoluble polymers. Stereoregular 1,5-substituted *hb*-PTAs, however, were successfully obtained when Cp*RuCl(PPh$_3$)$_2$ was used as the catalyst. Compared to their regiorandom congeners, completely soluble polymers *hb*-1,5-P(**30–20**) with similar isolation yields and molecular weights were obtained in much shorter polymerization times (Table 1, entries 3 and 4). One reason for the solubility difference in the hyperbranched PTAs might be because the triazole rings function as ligands that form complexes with copper but not ruthenium ions, which thus cross-links the *hb*-1,4-PTA spheres [55]. Another possible cause might be due to the difference in the architectural structures of 1,4- and 1,5-disubstituted triazole rings. The 1,4-isomer experiences little steric interaction with the neighboring phenyl groups and can arrange in a relatively planar confirmation (Fig. 1). In contrast, its 1,5-substituted congener experiences strong steric repulsion, forcing the molecule to adopt a more twisted conformation. The structural planarity of the 1,4-isomer may facilitate π–π stacking of the aromatic moieties, leading to a lower solubility.

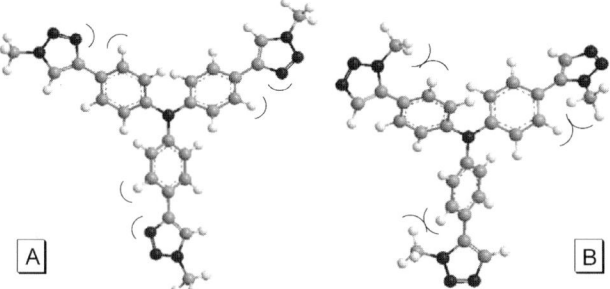

Fig. 1 Simulated conformations of **A** tris[4-(1-methyl-1*H*-1,2,3-triazol-4-yl)phenyl]amine and **B** tris[4-(1-methyl-1*H*-1,2,3-triazol-5-yl)phenyl]amine

Spectroscopic analysis revealed that the thermally initiated [3 + 2] polycycloaddition produced 1,4- and 1,5-substituted triazole isomers in an approximately 1 : 1 ratio. This ratio appears to be statistic and dependant on the bulkiness of the organic moieties. For example, hb-r-P[30(4)–20] with butyl spacers contained slightly more 1,4-triazole isomers than did hb-r-P[30(6)–20] with hexyl spacers. This becomes clearer if we look at the proposed transition states *a* and *b* of the [3 + 2]-dipolar cycloaddition (Scheme 16). Because of their molecular orbital symmetry, the acetylene and azide functional groups arrange in two parallel planes, a so-called "two-plane orientation complex" [48], which facilitates a concerted ring formation. If the monomer fragment or the polymer branch (∼∼∼) attached to the functional groups are bulky, steric repulsion will come into play and transition state *a* will be

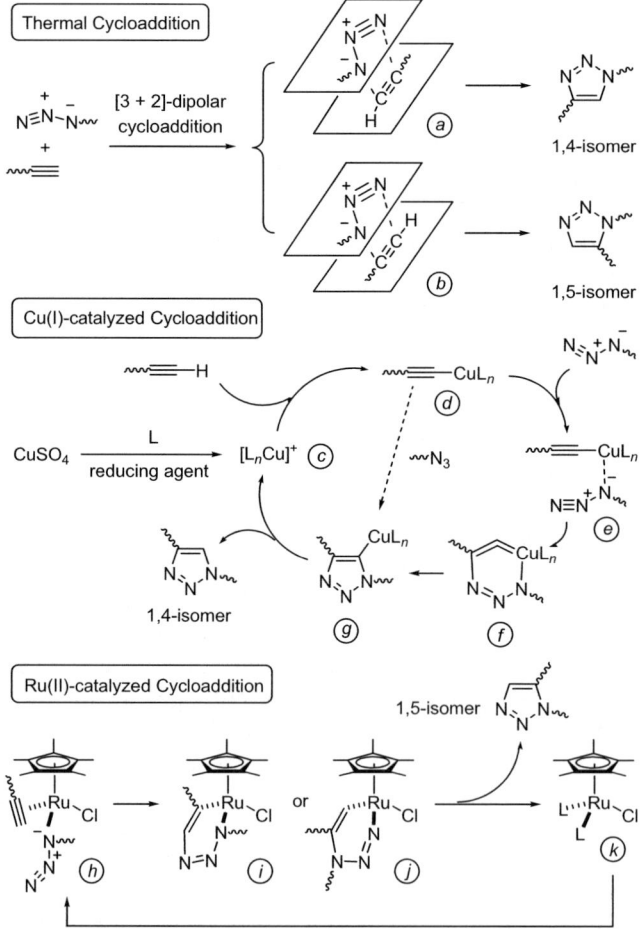

Scheme 16 Proposed mechanisms for thermal and catalytic cycloaddition reactions

favored, hence the formation of a hyperbranched polymer with a higher 1,4-isomer content. With a small steric interaction, both the two transition states are possible. As a result, the reaction gives a statistic mixture with both 1,4- and 1,5-isomeric structures.

Whilst the thermally induced 1,2,3-triazole formation between acetylene and azide yields a mixture of 1,4 and 1,5-isomers, stereoregularity can be achieved when the cycloaddition is catalyzed by a copper(I) salt (Scheme 16). The active catalytic species (c) is usually generated from a copper(II) salt in the presence of a reducing agent such as ascorbic acid or sodium ascorbate. The catalytic cycle commences with the generation of copper(I) acetylide (d). Density functional theory calculations show that a concerted cycloaddition directly from d to g is disfavored by ~ 12–15 kcal mol^{-1} and that the cycloaddition proceeds in a stepwise annealing sequence ($d \rightarrow e \rightarrow f \rightarrow g$), going through the intriguing species of six-membered metallocycle f [56]. As mentioned above, so far there has been no report in the literature utilizing this potential protocol to synthesize soluble 1,4-regioregular PTAs and the attempts made by others and us had all been unsuccessful. The mechanism of the ruthenium(II)-catalyzed synthesis of 1,5-substituted hb-1,5-PTAs is not well understood at present and needs more detailed investigations. The 1,5-disubstituted 1,2,3-triazole rings may have formed via an initial oxidative coupling of the alkyne and azide groups on the ruthenium metal center, followed by reductive elimination of the 1,5-isomeric unit from the six-membered ruthenacycles i and/or j (Scheme 16) [57]. The ruthenium(II) catalyst is also active for internal alkynes, forming 1,4,5-trisubstituted 1,2,3-triazoles in a regioselective manner [58]. This new protocol will help generate stereoregular trisubstituted hb-PTAs.

2.4
Polycyclotrimerization

All the hyperbranched polymers prepared by the synthetic routes discussed above have one feature in common: their branching units are already embedded in their monomer building blocks. In other words, nonlinear monomers with trifunctional groups are needed in order to construct the hyperbranched macromolecules. It is of great interest to develop new polymerization processes, where the propagation reaction inherently generates branching structures from simple, linear monomers. An example of such a reaction is transition metal-catalyzed polycyclotrimerization of diynes (Scheme 17). This kind of [2 + 2 + 2] cyclization reaction involves a single monomer species, suffers no stoichiometric constraint, and can potentially produce polymers with high molecular weights and high stability. Inherently the resultant polymers should be highly branched, because the polycyclotrimerization mechanism is intolerant of, or incompatible with, the formation of linear repeat units inside the hyperbranched cores.

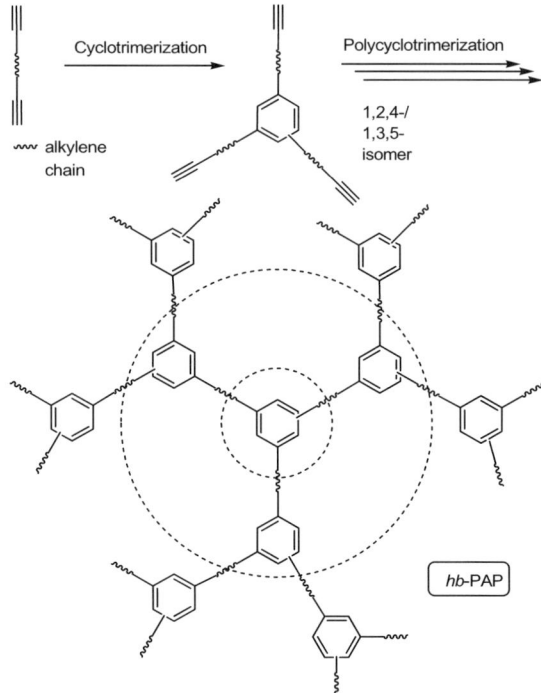

Scheme 17 Synthesis of hyperbranched polymer through diyne polycyclotrimerization

A few research groups had looked into the possibility of utilizing the alkyne cyclization for polymer synthesis in the early 1970s and late 1980s [59–63]. But as summarized by Sergeyev et al. in a symposium review article [59] at that time, the resultant homopolymers "are infusible and insoluble in conventional organic solvents" and "are difficult to reprocess into manufactured articles". The interest in the alkyne homopolycyclotrimerization had therefore subsided, with almost nobody revisiting the subject over the past decades, because the insolubility of the homopolymers of the diynes rendered their structural characterization a difficult proposition and their intractability made them almost useless in terms of finding practical applications as plastic materials.

2.4.1
Hyperbranched Poly(alkylenephenylene)s

We have studied homopolycyclotrimerizations of aliphatic terminal diynes with various alkylene spacer lengths catalyzed by tantalum and niobium halides or by binary mixtures of the metal halides and tetraphenyltin (Scheme 18) [64–66]. Under optimized conditions, completely soluble diyne homopolymers with high molecular weights (M_w up to $\sim 1.4 \times 10^6$) and pre-

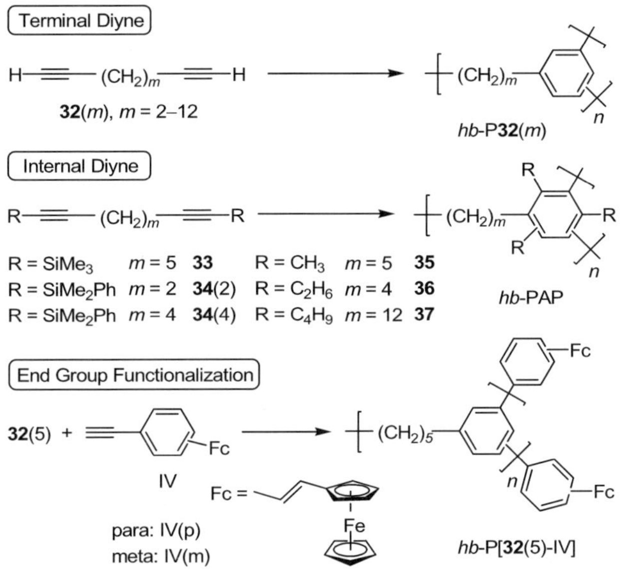

Scheme 18 Homopolycyclotrimerizations of terminal and internal diynes and synthesis of *hb*-PAPs with functional end groups

dominant 1,2,4-benzenetriyl core structures were obtained in high yields (up to 93%). Internal diynes could also be polymerized into hexasubstituted *hb*-PAPs [67]. Copolycyclotrimerization of the aliphatic diynes with monoynes enabled the incorporation of functional groups into the *hb*-PAP structure at the molecular level. This was demonstrated by the synthesis of *hb*-P[32(5)-IV]: the *hb*-PAP was decorated by redox-active ferrocene units on its periphery [68].

The alkyne cyclotrimerizations catalyzed by the complexes of late transition metals such as Ni, Zn, Rh, Pd, Ru, Co, Ir, etc. have been extensively studied and the involved reaction mechanisms have been well established [69–76]. Although the acetylene cyclotrimerizations catalyzed by the complexes of early transition metals have been less studied, highly regioselective systems have been developed [77–79]. For example, some titanium complexes have been found to catalyze the regioselective cyclotrimerizations of terminal alkynes in high efficiencies, giving 1,2,4-trisubstituted benzenes in excellent regioselectivities ($\geq 97\%$) and isolation yields ($\geq 95\%$) [77]. Group V transition metal complexes such as niobium and tantalum halides usually work very well and produce *hb*-PAPs with predominately 1,2,4-substituted benzenes. It has been proposed that tantallacyclic intermediates are involved in tantalum-catalyzed alkyne cyclotrimerizations [80–83]. The diyne polycyclotrimerization may have followed similar paths with the metallacyclic intermediates serving as the initiating and propagating species. An in-situ generated Ta(III) species [84, 85] may oxidatively add to a diyne (*a*) to form a tantallacyclopropene intermediate b (Scheme 19). Insertion of a triple bond of another diyne monomer to b gives three possible regioisomers of tantallacyclopentadienes c, among which c_I experiences the least steric repulsion from ligands R and may thus be preferentially formed. Addition of a third diyne to c_I can potentially form two tantallacycloheptatriene isomers (d) and two Diels–Alder adducts of tantallanorbornadienes (e). Each of the intermediates experiences different steric interactions but all of them give the same product of 1,2,4-trisubstituted benzene f through reductive elimination of the metal species. Following a similar pathway, 1,3,5-substituted benzene unit g may be generated from tantallacyclopentadiene c_{II}. The 1,2,4- and 1,3,5-arenes f and g formed during the initiation step will serve as propagating species to follow the same reaction pathway to grow to a higher generation. Iterative repeats of the oxidative addition–reductive elimination cycles in the propagation step eventually lead to the formation of an hb-PAP.

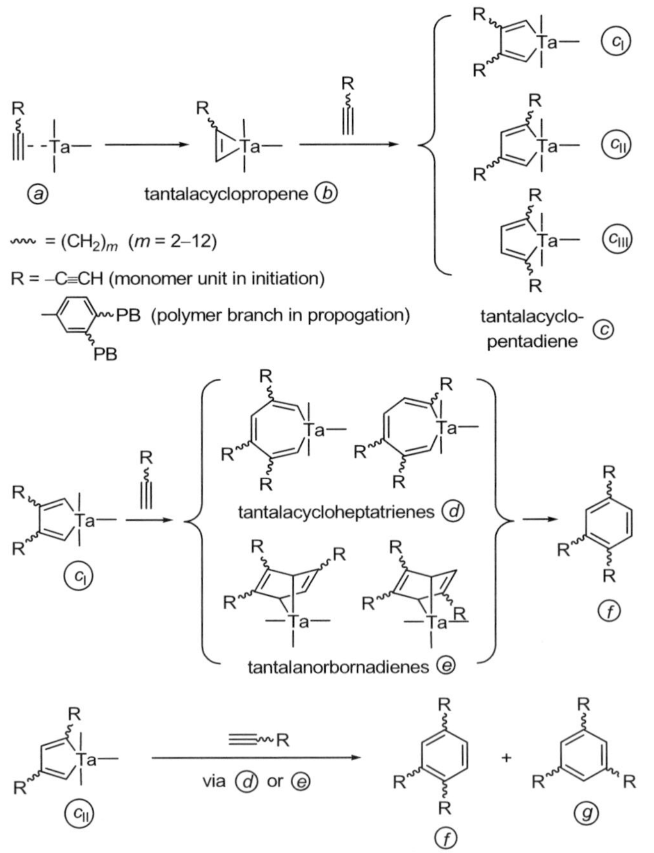

Scheme 19 Proposed mechanism for tantalum-catalyzed alkyne polycyclotrimerization

In Scheme 19, only is the possibility for one triple bond of a diyne monomer to react with a metal center considered. As outlined in Scheme 20, it is possible that through a "back-biting" reaction pathway, two triple bonds in one diyne molecule add to the same metal center to form a tantallacyclopentadiene intermediate c. Further reaction with another triple bond of a propagating branch would produce benzocycloalkene f, which will terminate the growth of the branch by end-capping it. A similar structure can be formed by the end-capping reaction of a propagating species d with diyne 32(m). The formation of tantallacyclopentadiene followed by the back-biting of the second alkyne easily leads to the ring closure due to their close proximity. Depending on the length of alkylene spacer, 4- to 8-membered benzannulated rings may be formed as end groups on the peripheries of hb-PAPs.

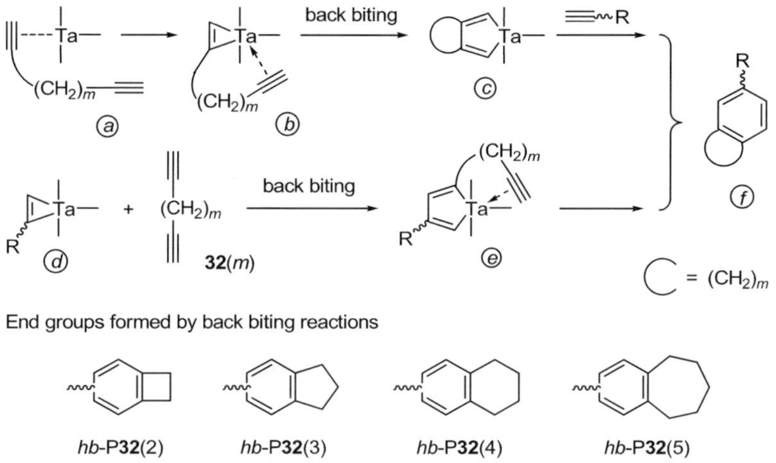

Scheme 20 Proposed mechanism for back-biting reactions occurring in polycyclotrimerization of aliphatic diynes

The back-biting reaction plays an important role in the polycyclotrimerization of the aliphatic diynes. If this end-capping reaction is too active, the propagating branches will be easily terminated, giving only oligomeric products. On the other hand, if the back-biting reaction is too sluggish, the polymerization will become difficult to control, resulting in the formation of cross-linked gels. Fine tuning the back-biting reaction will help control the formation of the reaction products. For example, diynes with two and three short methylene spacers such as 32(2) and 32(3), respectively, have a high tendency towards back-biting and thus form hb-PAPs of low molecular weights (Table 2, nos. 1 and 2). No terminal triple bonds but strong back-biting signals can be observed in their ^{1}H NMR analyses. Diynes with long spacers [32(m), m≥6] show low back-biting activity, which can be easily confirmed by the existence of their unterminated triple bonds. As a result, the polymer-

Table 2 Analysis of the products [hb-P32(m)] obtained from the polycyclotrimerizations of aliphatic diynes 32(m) with different lengths (m) of methylene spacers[a]

No.	Spacer	Solubility[b]	Triple bond[c]	Back biting[d]	M_w[e]	M_w/M_n[e]
1	$-(CH_2)_2-$	Soluble	Not observed	Strong	∼ 3000	2.3
2	$-(CH_2)_3-$	Soluble	Not observed	Very strong	∼ 900	1.5
3	$-(CH_2)_4-$	Soluble	Observed	Medium	∼ 60 000	4.9
4	$-(CH_2)_5-$	Soluble	Not observed	Medium	∼ 40 000	5.0
5	$-(CH_2)_6-$	Soluble	Observed	Very weak	∼ 600 000	23.0
6	$-(CH_2)_8-$	Insoluble				
7	$-(CH_2)_9-$	Soluble	Observed	Weak	∼ 200 000	7.9
8	$-(CH_2)_{10}-$	Insoluble				

[a] The polymerization reactions were carried out in toluene at room temperature under nitrogen by using TaCl$_5$ – Ph$_4$Sn as catalyst
[b] Tested at room temperature in common organic solvents including toluene, benzene, chloroform, DCM, and THF
[c] Signal in the ^{1}H NMR spectrum of hb-PAP
[d] Signal intensity in the ^{1}H NMR spectrum of hb-PAP
[e] Estimated by SEC in THF on the basis of a linear polystyrene calibration

izations are difficult to control, forming polymers with very high M_ws and extremely broad PDIs (Table 2, nos. 5 and 7) or even totally insoluble gels (Table 2, nos. 6 and 8).

A unique odd-even effect of the monomers in the diyne polycyclotrimerization was observed. In the aliphatic diynes with an odd number of methylene spacers, their triple bonds locate in the same side, which facilitates the back-biting reaction (Scheme 21). In contrast, in the diynes with an even number of methylene units, their triple bonds locate in the opposite sides: these unfavorable positions frustrate the back-biting reaction. Consequently, hb-P32(4) possessed triple bond residues in its final structure, whereas its congener hb-P32(5) did not.

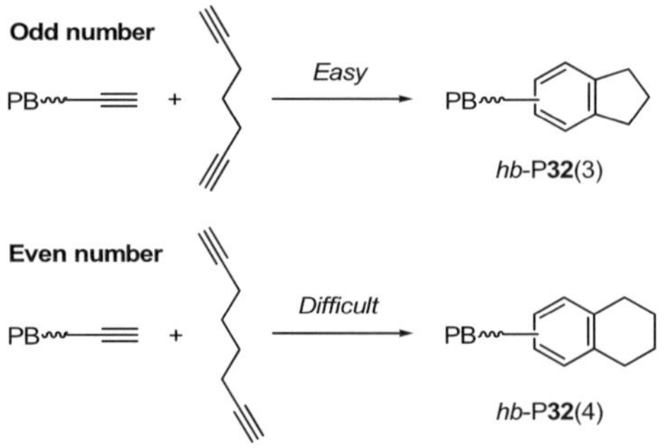

Scheme 21 Odd-even effect in the back-biting reaction

The hb-PAPs contain numerous branching units, resulting from the cyclization propagation and back-biting termination. In addition to the branching structures, there exist pseudo-"linear" structure in the polymers, formed by the reaction of the closely located triple bonds in a 1,2,4-substituted benzene ring (Scheme 22). The acetylene triple bonds in the 1 and 2 positions can form a new benzene ring through cyclotrimerization with another triple bond from a monomer or a polymer branch. Although this may also be considered as a "cross-linking" reaction, the structural motif is not detrimental to the solubility of hb-PAP due to its overall "linear" propagation mode. Combining all these three structural features, the hb-PAPs possess a molecular architecture similar to that of glycogen (Scheme 23), a hyperbranched biopolymer [66].

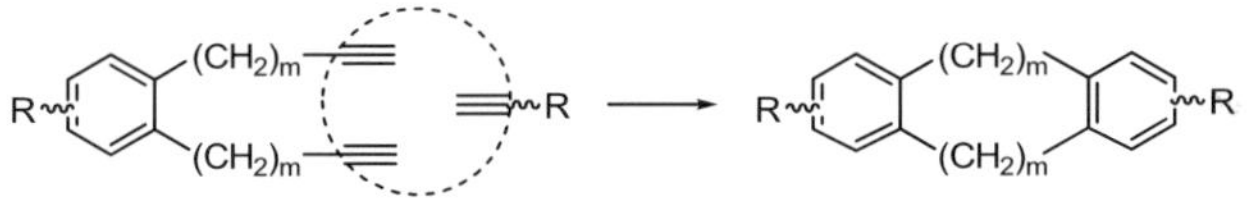

Scheme 22 Pseudo-"linear" propagation mode in the diyne polycyclotrimerization

Changing terminal diynes to internal ones changes the polymerizability of the diyne monomers as well as the structure of the resultant polymers. The attempted polymerizations of the internal diynes with such bulky groups as trimethylsilyl and dimethyl(phenyl)silyl groups either gave no polymeric products or produced oligomeric species (Table 3). Internal diynes with "slim" head groups such as Bu were active. Similar to the case of terminal diynes, if the alkyl spacer between the two triple bonds in the internal diynes was too long, the polymerizations of the monomers gave insoluble gels due to the absence of the growth-controlling back-biting termination. Only did

Table 3 Results of attempted polymerizations of internal diynes

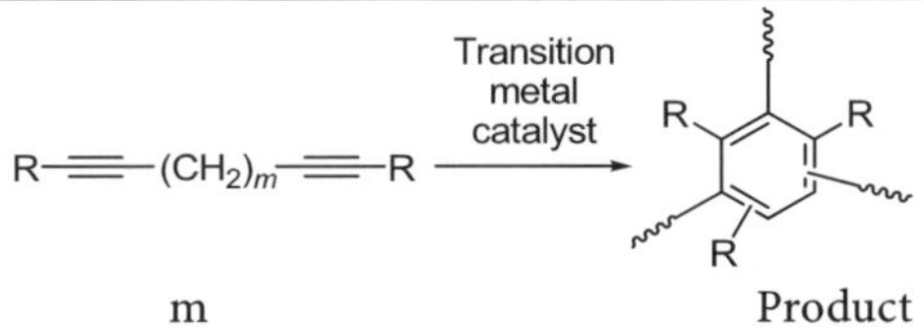

R	m	Product
– SiMe$_2$Ph	2, 4	No polymer or only oligomer
– SiMe$_3$	5	
– C$_4$H$_9$	12	Insoluble gel
– C$_2$H$_5$	4	Soluble polymer
– CH$_3$	5	

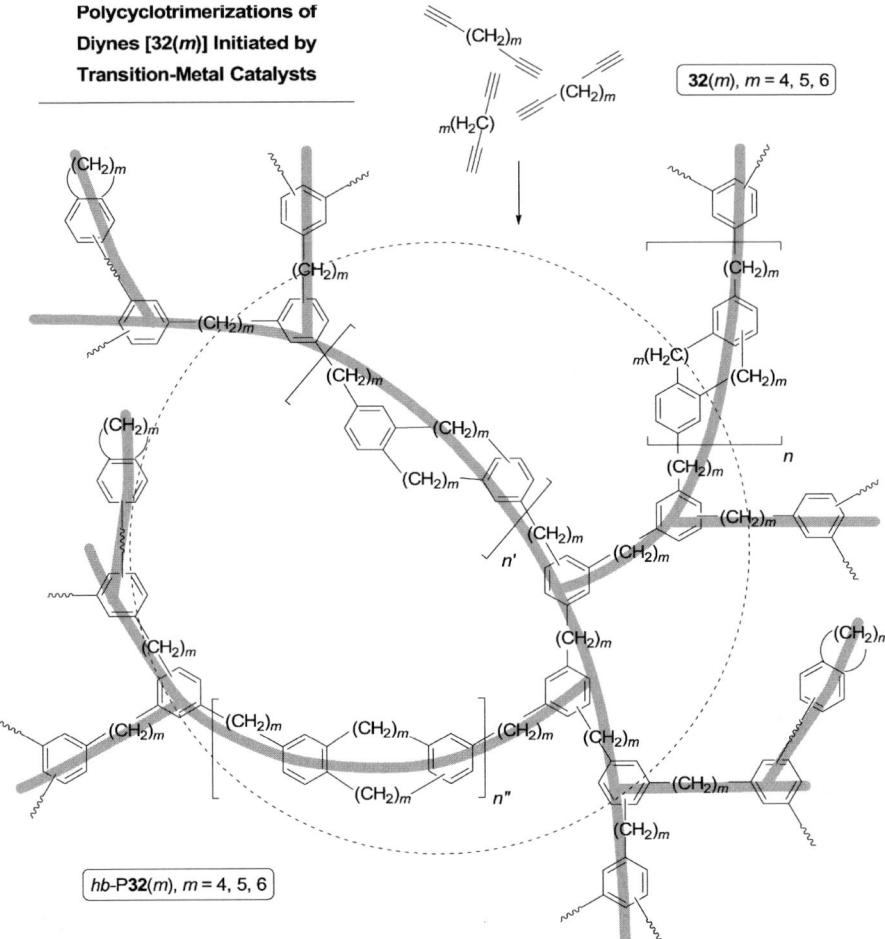

Scheme 23 Glycogen-like molecular architecture of *hb*-PAPs

those internal diynes with proper head groups and spacer lengths produce completely soluble, high molecular weight polymers.

2.4.2
Hyperbranched Polyarylenes

As discussed above, changing the diyne position from terminal to internal is one way to change the nature of the hyperbranched polymers. More profound changes can be brought about by replacing the aliphatic spacers with aryl rings. Different from the isolated benzene rings in the *hb*-PAPs, here the "new" benzene rings formed by the polycyclotrimerization help the "old" aromatic rings to interconnect into conjugated *hb*-PAs (Scheme 24).

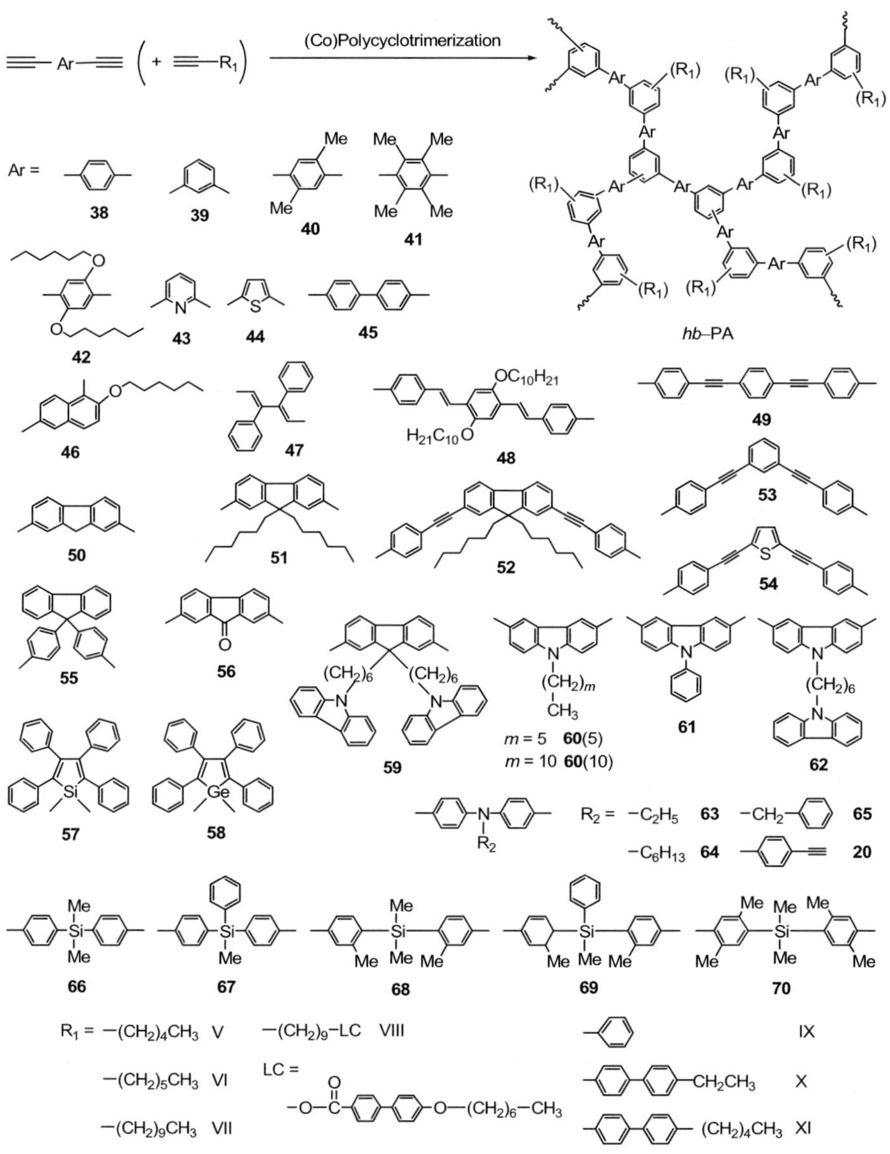

Scheme 24 Synthesis of *hb*-PAs by (co)polycyclotrimerization of aryldiynes (with monoynes)

Because of the lack of the back-biting capability of the rigid aromatic diynes, their homopolycyclotrimerizations are not as easy to control as those of their congeners of flexible aliphatic diynes. Nevertheless, high molecular weight *hb*-PAs could be obtained in quantitative yields from the homopolycyclotrimerizations of nitrogen-containing aromatic diyne (**59–65**) catalyzed by

CpCo(CO)$_2$ under UV irradiation and from the polymerizations of metallolyl (**57** and **58**) and silyldiynes (**66–70**) catalyzed by tantalum halides [86–91]. The homopolycyclotrimerizations of other aromatic diynes (**38–56**) all proceeded very rapidly, giving polymeric products that were only partially or totally insoluble in common organic solvents due to the involved cross-linking reactions. The large free volumes and irregular molecular structures generated by the nonlinear carbazolyl, diphenylamine, metallolyl and silyl groups may have helped endow the homopolymers (*hb*-P57 to *hb*-P70) with the excellent solubility.

Structural analysis of the homopolymers by spectroscopic methods confirmed that the diynes had undergone [2 + 2 + 2] polycyclotrimerizations by forming new benzene rings from their acetylenic triple bonds. The ratio of the 1,2,4- to 1,3,5-isomers of the trisubstituted benzene rings was estimated to be $\sim 2.2 : 1$. Careful evaluation of the ^{1}H NMR spectra unveiled that the number of terminal triple bonds in the final *hb*-PAs was much smaller than that in an "ideal" hyperbranched structure produced by the diyne polycyclotrimerization. This result suggests that intra-sphere ring formation might have been involved in the cyclotrimerization polymerization.

To account for the "missing" acetylenic protons, three possible pathways for the consumption of the triple bonds were proposed and are depicted in Scheme 25. The first pathway is via a normal P$_1$M$_2$ growth mode, where P and M stand for polymer and monomer, respectively. The second is via an intracyclotrimerization mode of P$_2$M$_1$ type, with two triple bonds from two polymer branches and one from a monomer involved. The third is via another "pure" intracyclotrimerization mode of P$_3$ type, with the three triple bonds all from the branches of a polymer. The experimental results indicate that one or both of the intracyclotrimerization modes must have been at play in the polycyclotrimerization of silylenediynes. However, the newly formed benzene rings by these intracyclotrimerizations are undistinguishable from each other and are also indistinct from those formed by the normal polycyclotrimerization reactions in the NMR spectra, making it difficult to experimentally determine the probabilities of the intracyclotrimerization reactions.

To solve this problem, computational simulation was exercised. The models of the polymers were built, and the probabilities of the growth modes were estimated according to the minimized energy of the structures, using the Materials Studio program[1]. An example of the outputs of the computer simulations is shown in Fig. 2. The overall structure of *hb*-P66 looks like a star-shaped macromolecule containing a number of small cyclic units (Fig. 2, lower right panel). The total number of the triple bonds left inside the hyperbranched structure and the total number of the aromatic protons were in agreement with the numbers estimated from the ^{1}H NMR analysis.

---

[1] Materials Studio is a software environment that brings together the world's most advanced materials simulation and informatics technology. It is a product of Accelrys Inc.

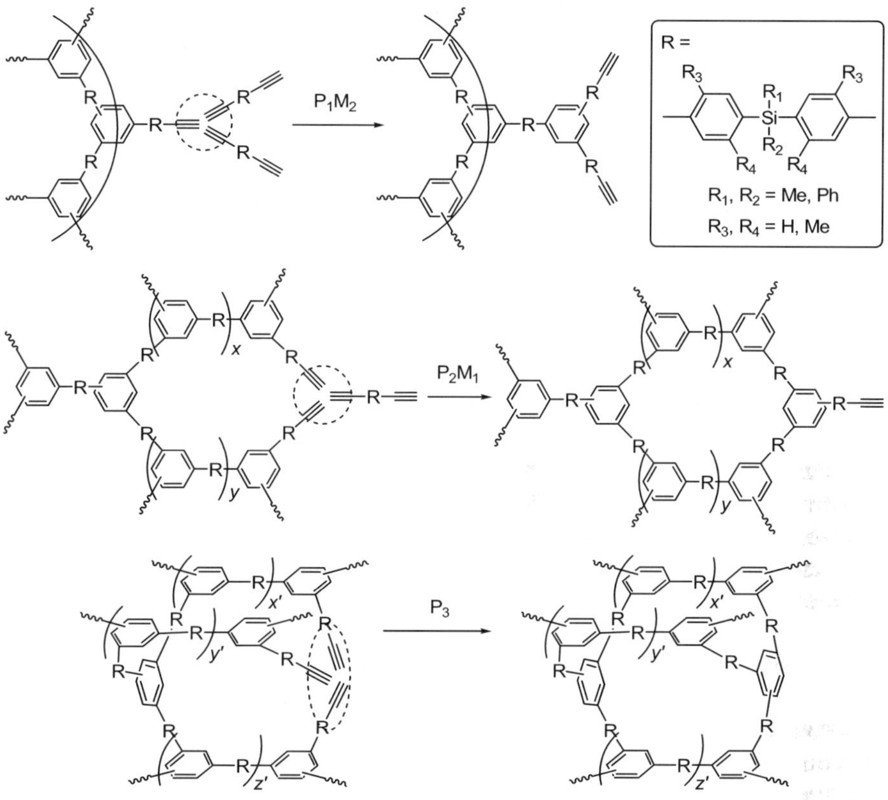

Scheme 25 Propagation modes for the homopolycyclotrimerization of silyldiynes

The computer simulation model is thus consistent with the structure of the real polymer. Estimation of the three different growth modes gave possibilities of 64%, 34% and 2% for the P_1M_2, P_2M_1 and P_3 modes, respectively. Although more than 1/3 (or 36%) of the propagation modes are via the intracyclotrimerization reactions, most of the cycles are small, being formed by only two monomer units mainly due to the close proximity of the 1 and 2 positions of the newly formed 1,2,4-trisubstituted benzenes. The small rings are strung together like beads in a necklace. This structure model is in agreement with the excellent solubility of the polymer, although it contains numerous cyclic structures.

As mentioned above, the homopolycyclotrimerization was limited to a small number of diynes in terms of generation of soluble polymers. To overcome the problem of uncontrolled cross-linking reactions and to improve the solubility of the polymers, copolycyclotrimerizations of aromatic diynes with monoynes (V–XI) were carried out (cf., Scheme 24). This approach worked very well: all the copolycyclotrimerization reactions proceeded smoothly with good controllability, producing completely soluble hyperbranched copoly-

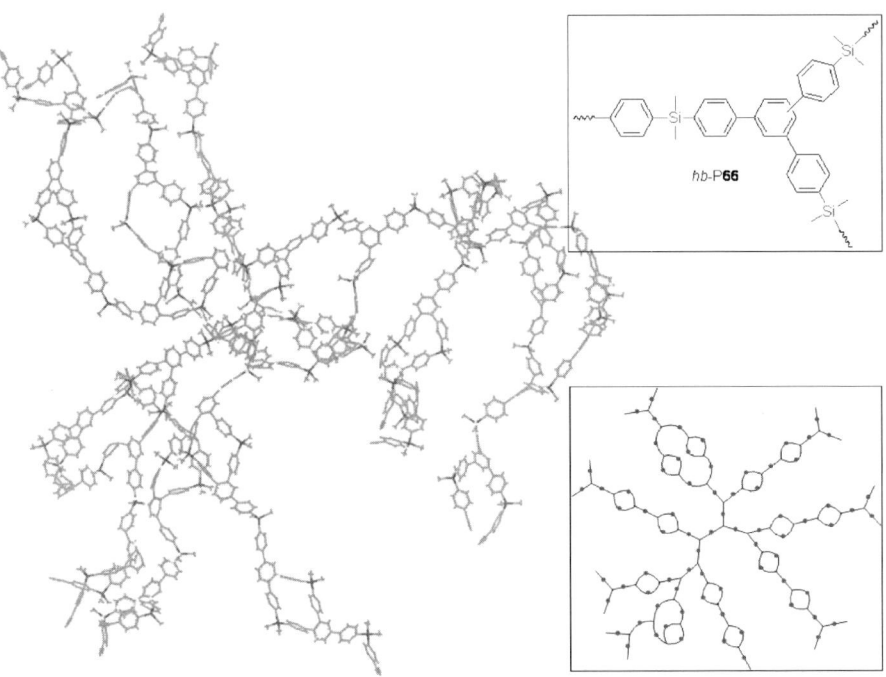

Fig. 2 Three-dimensional macromolecular structure of *hb*-P66 simulated by Materials Studio program. *Inset*: Chemical structure of *hb*-P66 (*upper panel*) and simplified illustration of the two-dimensional geometric structure of *hb*-P66 (*lower panel*)

arylenes with high molecular weights (M_w up to $\sim 1.8 \times 10^5$) in high isolation yields (up to 99.7%) [88, 92–98]. 1-Alkynes (V–VIII) are generally better comonomers than 1-arylacetylenes (IX–XI) when the molecular weights of the *hb*-PAs are concerned. This is probably due to two effects. First, the long alkyl chains may confer higher solubility on the propagating species, therefore enabling their continued, further growth into bigger polymers. The second might be associated with an electronic effect. The electron-donating alkyl groups make the triple bonds of 1-alkynes electronically richer, which are likely to favorably interact with the electron-poorer aromatic diynes [99], hence promoting the formation of high molecular weight *hb*-PAs.

As a nice example, a "true" *hb*-PP without any substituent groups (or a "pure", all-aromatic plastic) was readily synthesized by the copolycyclotrimerization of diethynylbenzene (**38**) with phenylacetylene (**IX**) (Scheme 26). Whilst its cousin of linear PPP becomes insoluble and intractable when the molecular weight of PPP reaches just a few thousands, the highly branched and irregular molecular structure of *hb*-P(**38**-X) hampers strong π–π stacking of the aromatic rings, making the polymer completely soluble in common organic solvents.

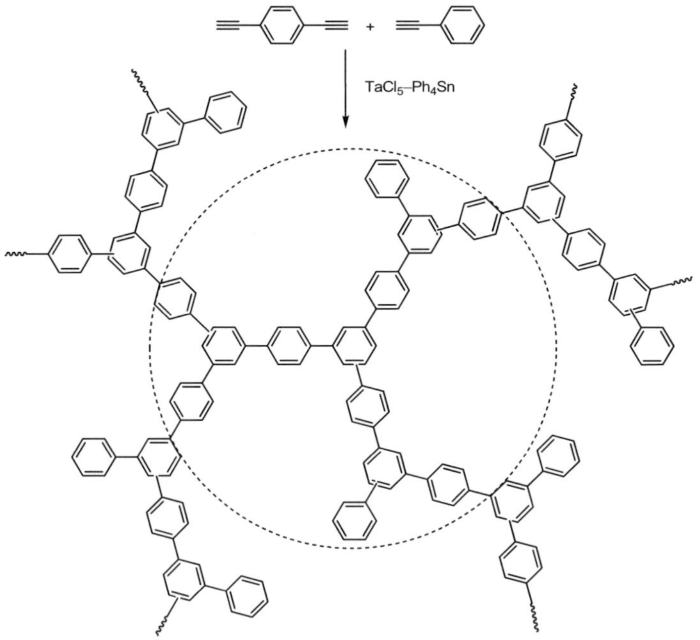

Scheme 26 Synthesis of *hb*-PP through copolycyclotrimerization of aryldiyne with arylmonoyne

Spectral characterization confirmed the proposed hyperbranched architecture of the copolyarylenes. Estimations of the ratio of diyne to monoyne incorporated into the *hb*-PAs revealed that the monoynes functioned as the growth-controlling agents, impeding intracyclotrimerization reactions [86]. As an optimized ratio of diyne to monoyne, 1 : 1.5 has often been found to work well to yield completely soluble, high molecular weight *hb*-PAs. Copolycyclotrimerization of aromatic triynes such as **20** is inherently much more difficult to control. While for aliphatic monoynes like 1-octyne (**VI**) a monoyne to triyne ratio of 3 : 1 was sufficient to obtain a soluble *hb*-PA, a larger ratio (4 : 1) was necessary for aromatic monoynes such as phenylacetylene (**IX**) [86, 100].

Detailed structural investigations revealed that even when such a large excess of growth-controlling monoyne was used, the resultant polymers still contained internal cyclic structures. According to the ratio of monoyne to triyne found in *hb*-P(**20**-**VI**), different propagation modes accounting for the intracyclotrimerization reactions were proposed as depicted in Scheme 27. Taking the 1,2,4- and 1,3,5-isomeric structures of the trisubstituted benzenes into account, the formation of "small"- and "medium"-sized cycles is highly possible. Similar to the silyldiyne homopolycyclotrimerization, the close intramolecular proximity of the two triple bonds originating from an ortho-connection are likely to be terminated by the triple bond of 1-octyne,

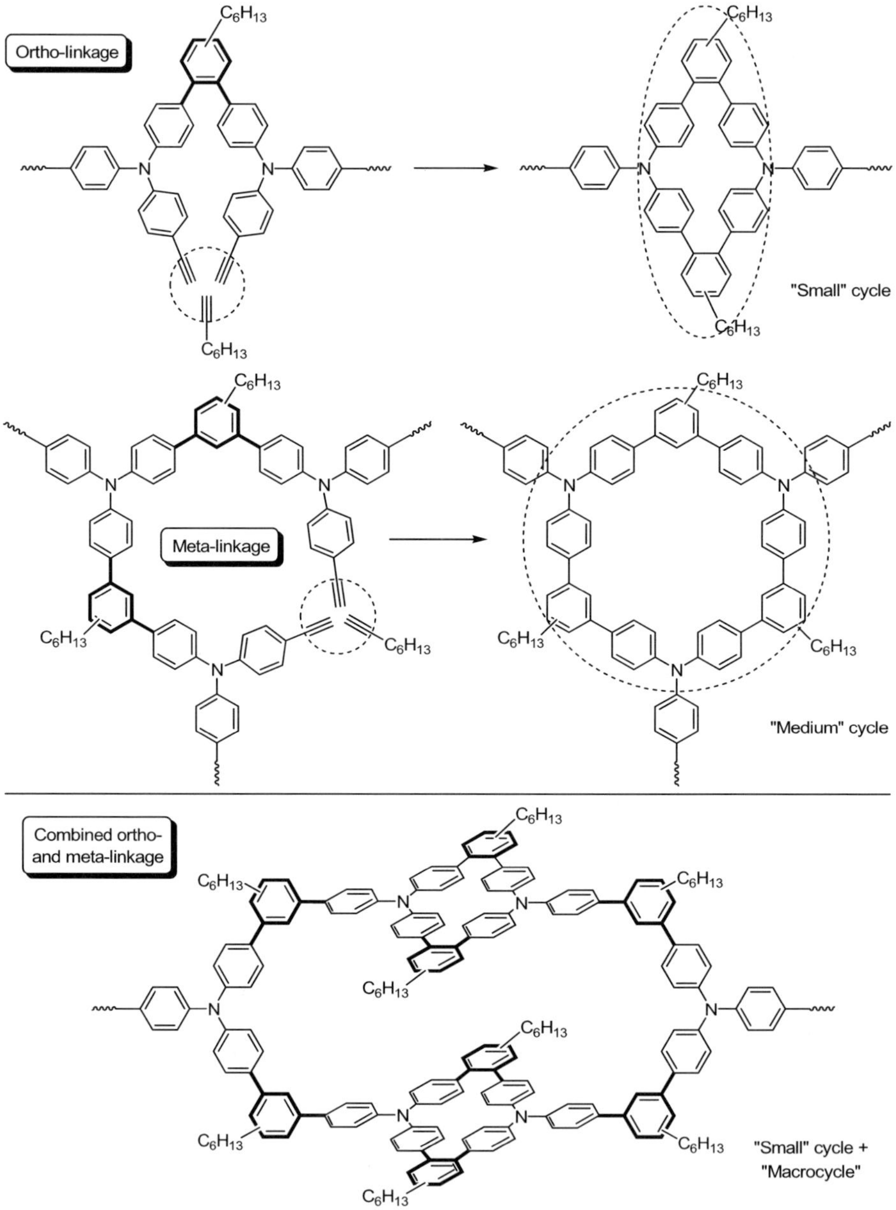

Scheme 27 Formation of intramolecular cyclic structures in alkyne copolycyclotrimerization

thus furnishing a second 1,2,4-trisubstituted benzene ring. This intramolecular "cross-linking" leads to the formation of a small cycle, yet the polymer is still soluble due to the overall pseudo-"linear" propagation mode.

Another possible pathway is the ring closure of the three triyne monomer units connected in a meta fashion with a monoyne via 1,2,4- or 1,3,5-reaction. Such a reaction is likely to produce a medium-sized ring, serving as a core for the macro-dentritic propagation with three growing arms. Para-substituted polymer branches resulting from 1,2,4-cyclotrimerization inherently cannot form any ring structures and may only be involved in the formation of oval-shaped "macrocycles". Such cyclic substructures are possibly formed via combined ortho- and meta-linkages, which exerts little effect on the solubility of the hb-PAs, as the overall structure is still propagating in a pseudo-"linear" mode.

2.4.3
Hyperbranched Poly(aroylarylene)s

In the previous section, we have described our successful syntheses of the soluble hyperbranched polymers by the [2 + 2 + 2] (co)polycyclotrimerizations of alkynes initiated by transition metal catalysts. All of the (co)polymers, however, possess regioirregular structures originating from the isomeric structures of 1,3,5- and 1,2,4-trisubstituted benzene rings. This structural irregularity is not necessarily a disadvantage and can actually be beneficial for the solubility and processability of the hyperbranched polymers. On the other hand, it makes it a challenging job to accurately characterize the molecular structures of the polymers by spectroscopic methods. Furthermore, the transition metal catalysts are highly moisture-sensitive and have little tolerance to polar functional groups [93]. The cyclotrimerization of monoynes of benzoylacetylenes catalyzed by the secondary amines is known to proceed in a strictly 1,3,5-regioselective fashion due to the ionic mechanism of the reaction [101–103]. The reaction proceeds in the absence of metallic catalysts: this metal-free feature makes the reaction attractive to us. We therefore explored the possibility of utilizing this cyclotrimerization reaction to prepare hb-PAAs from aroyldiynes bearing functional groups. We designed and synthesized a series of new bis(aryl ethynyl ketone)s with various organic and organometallic linkers and investigated their polymerization behaviors (Scheme 28) [104, 105].

It is known that an aroylacetylene undergoes cyclotrimerization in the presence of diethylamine or when refluxed in DMF [101–103]. Supposedly, a small amount of DMF solvent has decomposed at the high temperature (i.e., the boiling point of DMF) to release dimethylamine, which served as the catalytic species [103]. Our attempted polymerization carried out in the presence of diethylamine, however, produced polymers in very low yields (7–21%) [105]. Good to excellent polymer yields were achieved when the bis(aroylacetylene)s were refluxed in DMF/tetralin mixtures for 72 h. Our attempts to shorten the reaction time and to increase the polymer yield by using diphenylamine as the catalyst failed. Use of piperidine as the cata-

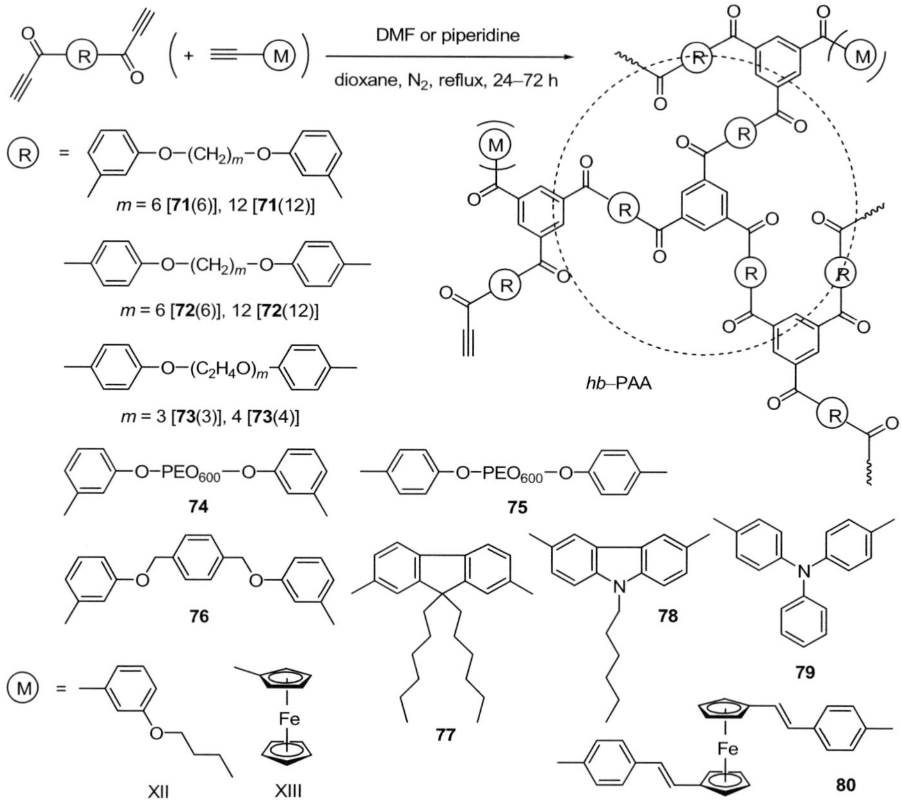

Scheme 28 1,3,5-Regioselective polycyclotrimerization of aroylarylenes

lyst, however, furnished soluble polymers in much shorter reaction times and higher yields: for example, hb-P71(6) and hb-P73(4) were obtained in virtually quantitative yields after 24 h reaction.

As proven by the spectroscopic analyses, the aroyldiynes were regioselectively polycyclotrimerized into hb-PAAs by piperidine. This regioselectivity stems from the ionic mechanism of the base-catalyzed polycyclotrimerization reaction [106]. Piperidine may have reacted with an aroylethynyl group in a Michael addition mode to form a ketoenamine (a), which further reacts with two more aroylethynyl groups to give a dihydrobenzene (d; Scheme 29). The piperidine moiety of d is removed by its reaction with another aroylethynyl group and aromatization gives a 1,3,5-trisubstituted benzene ring e. Repeats of this cycle result in the formation of an hb-PAA [105].

Again, with the help of spectral analyses, the hyperbranched structures of the poly(aroylarylene)s were confirmed. Thanks to the 1,3,5-regioselectivity of the cyclotrimerization reaction, the NMR spectra of the hb-PAAs were

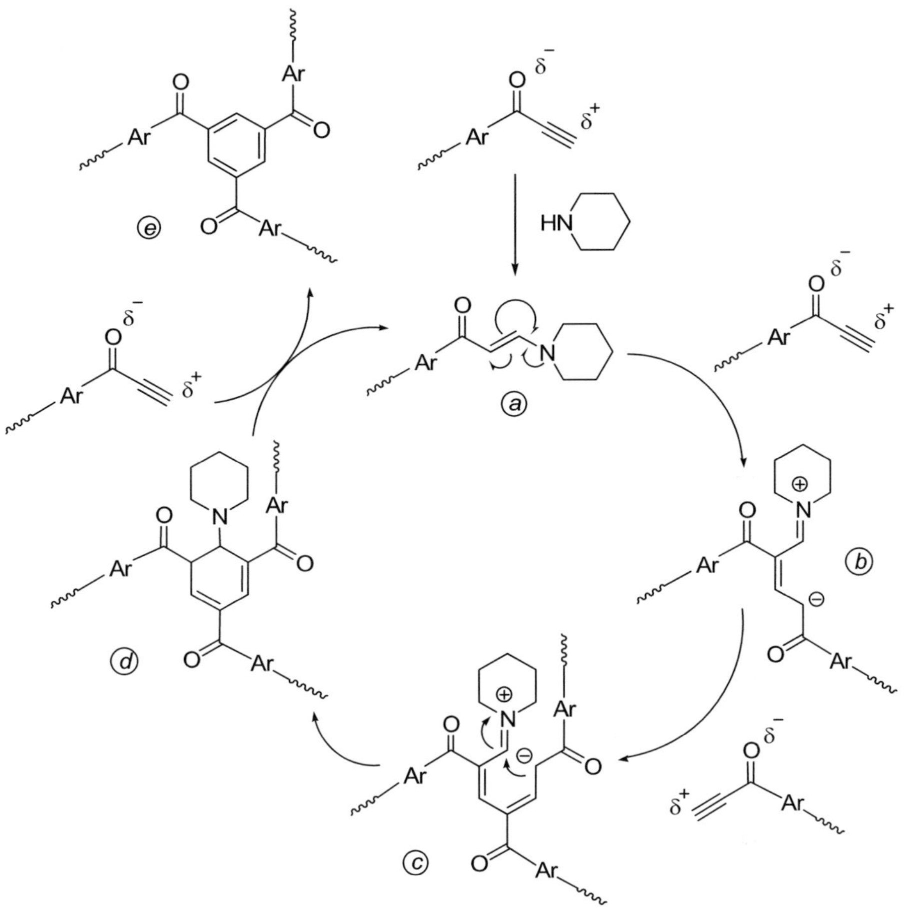

Scheme 29 Proposed mechanism for 1,3,5-regioselective polycyclotrimerization of aryl ethynyl ketone [102, 103]

much simpler, in comparison to their *hb*-PA congeners. The presence of three basic structures of a hyperbranched polymer, viz., D, L and T unites, was readily verified by the number of reacted acetylene triple bonds in the repeat unit (Scheme 30). Because of the structural flexibility of the monomers, especially **71–76**, the polymer branches can also be terminated by an additional dimerization end-capping reaction, in which two alkyne groups of the same diyne molecule react with a ketoenamine to form a new benzene ring. The end group resulting from this back-biting reaction is called "cyclophanic terminal" (T_c). From their corresponding resonance peaks in the ^{1}H NMR spectra, DB values of 78–100% were estimated. These values are much higher than those of the "conventional" hyperbranched polymers, which are commonly in the neighborhood of 50% [107].

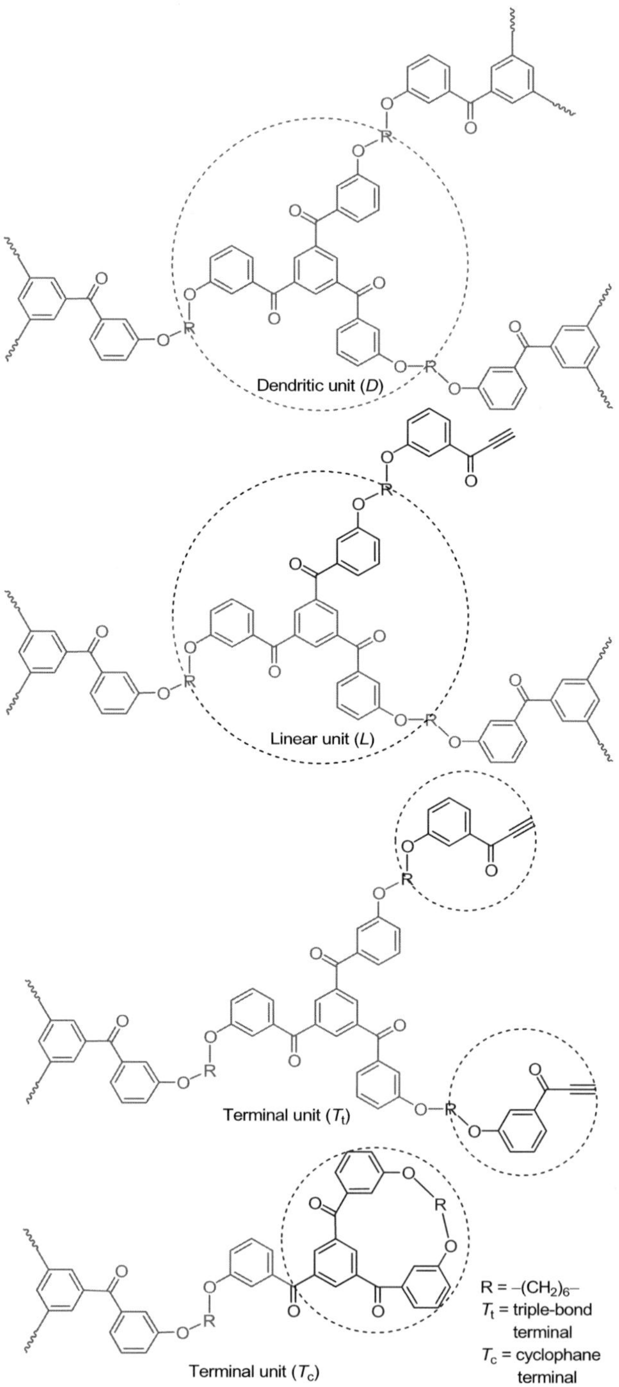
Dendritic unit (D)
Linear unit (L)
Terminal unit (T_t)
Terminal unit (T_c)
R = –(CH$_2$)$_6$–
T_t = triple-bond terminal
T_c = cyclophane terminal

Scheme 30 Molecular structures of *D*, *L* and *T* units of *hb*-PAA

3 Properties

All the herein-described hyperbranched polymers are constructed from triple-bond building blocks and are expected to show unique properties due to their novel π-conjugated structures. Investigation and understanding of the advanced functional properties of the new polymers will aid their development from *macromolecules* of academic curiosity to *materials* of technological value.

3.1 Thermal and Optical Properties

The hyperbranched polymers are carbon-rich macromolecules and show excellent thermal stabilities. The thermal properties of the *hb*-PAs are described below as an example. Their thermal stabilities were evaluated by TGA. Figure 3 shows TGA thermograms of some *hb*-PAs and Table 4 lists their thermal analysis data. The *hb*-PAs were thermally very stable: for instance, *hb*-P66 lost merely 5% of its weight at a temperature as high as 595 °C. All the polymers, except for *hb*-P(44-VI) and *hb*-P(59-VI), carbonized in > 50% yields on pyrolysis at 800 °C, with *hb*-P(45-V) graphitized in a yield as high as 86% (Table 4, no. 3). The thermal stabilities of the *hb*-PAs are similar to that of linear polyarylenes such as PPP but different from those of linear polyacetylenes such as PH and PPA. The dramatic difference in the thermal stability is mainly due to the structural difference: PPP is composed of thermally stable aromatic rings ($T_d \sim 550\,°C$) [108–112], whereas PPA and PH are comprised of labile polyene chains, which start to decompose at temperatures as low as 220 and 150 °C, respectively [113]. The excellent thermal stabilities of the *hb*-PAs

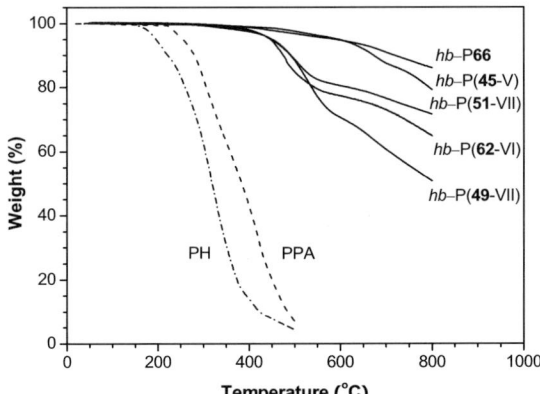

Fig. 3 TGA thermograms of *hb*-PAs; data for linear PH and PPA are shown for comparison

Table 4 Thermal and optical properties of hb-PAs

No.	hb-PA	T_d [a]	W_r [b]	λ_{em} [c]	Φ_F [d]	T_L [e]	F_{OL} [f]	$F_{t,m}/F_{i,m}$ [g]
1	P(**38**-VI)	452	71[h]	400	94	59	1016	0.15
2	P(**44**-VI)	440	0	486	14	48	802	0.13
3	P(**45**-V)	585	86	398	49	58	343	0.21
4	P(**45**-VI)	463	66	400	74	69	1265	0.15
5	P(**45**-X)	412	50	397	9	92	900	0.52
6	P(**49**-VII)	449	51	399	46	64	260	0.11
7	P(**50**-V)	467	75	400	31	66	126	0.11
8	P(**50**-VI)	451	70	400	86	48	1000	0.15
9	P(**50**-VII)	459	65	400	98	65	509	0.11
10	P(**52**-VII)	458	53	399	15	44	155	0.08
11	P(**59**-VI)	414	16	402	90	72	2300	0.63
12	P[**60**(5)-V]	487	58	400	7	84	1400	0.36
13	P[**60**(10)-V]	404	83	398	10	83	635	0.36
14	P[**60**(10)-IIX]	474	64	399	20	83	577	0.35
15	P(**61**-VI)	477	70	398	21	49	1034	0.19
16	P(**62**-VI)	463	65	396	28	46	1050	0.17
17	P**66**	595	79	402	1	85	1500	0.43

[a] Temperature (°C) for 5% weight loss
[b] Weight of the residue (%) left at 800 °C unless otherwise specified
[c] Peak of emission spectrum (nm) in DCM solution
[d] Fluorescence quantum yield (%) estimated by using 9,10-diphenylanthracene (Φ_F = 90% in cyclohexane) as standard
[e] Linear transmittance (%)
[f] Optical limiting threshold (mJ/cm^2) defined as the incident fluence at which the nonlinear transmittance is 50% of the initial linear one
[g] Signal suppression (ratio of saturated transmitted fluence to maximum incident fluence)
[h] At 750 °C

thus support the structural analyses by the spectroscopic methods, verifying their hyperbranched polyarylene structures composed of stable aromatic rings instead of labile polyene chains.

Organic molecules containing diyne units readily react upon heating [114–117] and many oligomers or prepolymers with monoyne end groups have been easily converted into thermoset networks [118]. The homo- and copolyynes containing both mono- and diyne moieties are thus expected to undergo facile thermal curing reactions. When hb-P(**19**-I) was heated in a DSC cell, it started to release heat at ca. 200 °C due to the commencement of thermally induced cross-linking reactions (Fig. 4). The exothermic reaction peaked at around 270 °C. The second heating scan of the DSC analyses gave almost a flat line parallel to the abscissa in the same temperature region, suggesting that all the acetylene triple bonds have reacted during the first heating scan. The cross-linking reaction of hb-P**20** started from ∼ 150 °C and peaks at ∼ 204 °C. Generally, the homopolyynes commenced to cure at lower tempera-

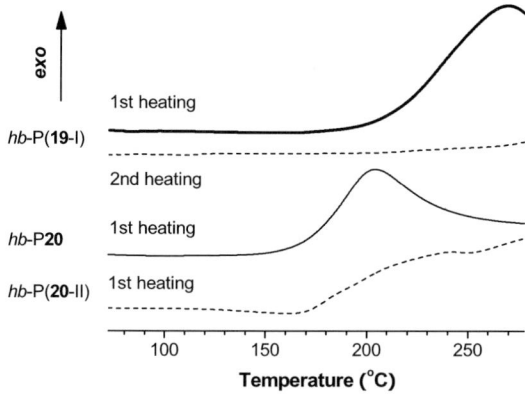

Fig. 4 DSC thermograms of *hb*-PYs measured at a scan rate of 10 °C/min under nitrogen

tures in comparison to their copolyyne congeners, which is probably because the former have more reactive terminal acetylene peripheries [119] and sterically less crowded aryl cores. When the terminal acetylene groups of *hb*-P20 were end-capped by phenyl groups, the resultant *hb*-P(20-II) contained only internal acetylene groups, which needed higher temperatures to initiate and complete its thermal curing reactions. This further manifests the effect of the acetylene reactivity on the thermal curability of the *hb*-PYs.

Many of the hyperbranched polymers contain aromatic chromophores in their π-conjugated structures and should show interesting optical properties [31]. This is indeed the case. For example, *hb*-P20 carries the TPA chromophore, which is often used as hole-transport materials in the fabrication of light-emitting diodes [120]. The λ_{ab} of triyne **20** appeared at 342 nm (Fig. 5),

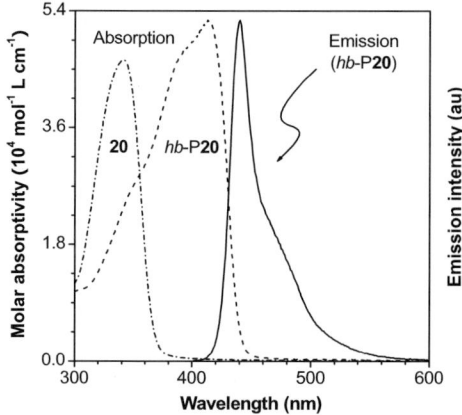

Fig. 5 Absorption spectra of THF solutions of triyne **20** and its polymer *hb*-P20 ($c = 12\,\mu g/mL$) and emission spectrum of the *hb*-P20 solution ($\lambda_{ex} = 368$ nm)

which was 43 nm red-shifted from that of TPA [121], thanks to the electronic communication of the peripheral triple bonds with TPA core. The λ_{ab} of hb-P20 was 413 nm, which was further bathochromically shifted from that of **20** by 71 nm, indicative of an extensive π-conjugation in the hb-PY. The polymer emitted a blue light of 440 nm upon photoexcitation. The emission was very bright, whose luminance easily goes beyond 1000 cd/m^2 when the polymer was excited by a weak UV lamp with a power of merely 30 mW/cm^2. The absorption and emission spectra of copolymer hb-P(**20**-I) resembled those of its homopolymer counterpart hb-P**20**, suggesting that the monoyne comonomer has exerted little effect on the electronic transitions of the copolymer.

Other hyperbranched polymers showed similar absorption and luminescence properties. Upon photoexcitation, the hb-PA solutions emitted deep-blue to blue-green lights, whose intensities were higher than that of poly(1-phenyl-1-octyne), a well-known highly emissive polyene. The PL efficiencies of the polymers varied with their molecular structures. Polymers hb-P(**38**-VI), hb-P(**45**-V), hb-P(**48**-VI), hb-P(**50**-VI), hb-P(**50**-VII) and hb-P(**59**-VI) exhibited Φ_F values higher than 70%, with hb-P(**50**-VII) giving the highest Φ_F value of 98%.

During our search for efficient light-emitting materials, we discovered a group of highly emissive molecules of metalloles such as silole (**57**) and germole (**58**). We synthesized hb-PAs containing the metallole moieties and observed a unique phenomenon of cooling-enhanced light emission (Fig. 6) [90]. When a solution of hb-P(**57**-VI) was cooled, its PL intensity was dramatically increased. At the low temperatures, the intramolecular rotations of the phenyl rings round the axes of the metallole core were impeded, which efficiently blocked the non-irradiative decay pathways [122]. A similar phenomenon was observed in the system of its congener, hb-P(**58**-VI) [89].

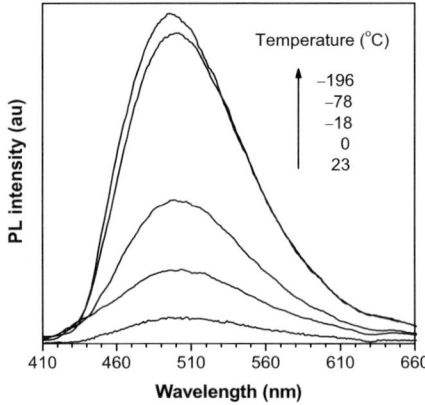

Fig. 6 PL spectra of silole-containing polymer hb-P(**57**-VI) in dioxane at different temperatures (c = 10 μM, λ_{ex} = 407 nm)

Figure 7 shows examples of nonlinear attenuation of the optical power of 532 nm laser pulses by hb-PA solutions. The transmitted fluence of hb-P(**49**-VII) initially increased with an increase in the incident fluence in a linear fashion. It started to deviate from the linearity at an incident fluence of ~ 260 mJ/cm^2 and reached a saturation plateau of 140 mJ/cm^2. The optical limiting performance of hb-P(**49**-VII) is superior to that of C$_{60}$, a well-known optical limiter [123]. The three-dimensionally conjugated structure of the hb-PAs may have been responsible for their optical nonlinearity, taking into account that C$_{60}$ is a three-dimensionally conjugated buckyball. In comparison to hb-P(**49**-VII), hb-P(**52**-VII) was a better optical limiter (maximum transmitted fluence < 100 mJ/cm^2) but P(**50**-VI) was a poorer one [93]. Clearly, the optical power-limiting properties of the hb-PAs are sensitive to the changes in their molecular structures, which offers opportunities to tune their NLO properties through molecular engineering endeavors.

Second-order NLO-active molecules have attracted much attention due to their attractive photonic applications [124–126]. A major effort in the area has been to efficiently translate large β value into high SHG coefficient d_{33}. The greatest obstacle has been the chromophoric aggregation in the thin films, which often quenches the EO activity in the solid state [127–129]. The chromophoric units of the NLO dyes are usually highly polarized by the push-pull interactions. During film fabrication, the chromophores with large dipole moments tend to compactly pack owing to the strong intermolecular electrostatic interactions, leading to the diminishment or cancellation of the

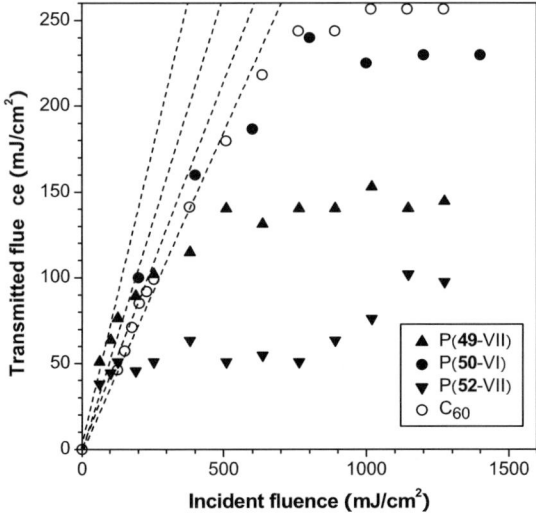

Fig. 7 Optical limiting responses to 8 ns, 532 nm optical pulses, of DCM solutions (0.86 mg/mL) of hb-PAs. Data for a toluene solution of C$_{60}$ (0.16 mg/mL) is shown for comparison

NLO effects in the thin films. Hyperbranched polymers should be ideal matrix materials as they offer three-dimensional spatial separation of the NLO chromophores in the spherical architecture, and their void-rich topological structure should help minimize optical loss in the NLO process.

The *hb*-PAEs of *hb*-P13 and *hb*-P15 contain NLO-active azo-functionalities, which are soluble, film-forming, and morphologically stable ($T_g > 180\,°C$). Their poled films exhibited high SHG coefficients (d_{33} up to 177 pm/V), thanks to the chromophore-separation and site-isolation effects of the hyperbranched structures of the polymers in the three-dimensional space (Table 5) [28]. The optical nonlinearities of the poled films of the polymers are thermally stable with no drop in d_{33} observable when heated to 152 °C (Fig. 8), due to the facile cross-linking of the multiple acetylenic triple bonds in the *hb*-PAEs at moderate temperatures (e.g., 88 °C).

Advanced photonic devices are often composed of working units with high RI contrasts. The RI values of existing polymers, however, vary in a small range ($n = 1.338–1.710$) [130, 131], which limits the scope of their photonic applications. Theory predicts that molecules consisting of groups with high

Table 5 NLO properties of *hb*-PAEs

hb-PAE	l_f (μm)[a]	d_{33} (pm/V)[b]
hb-P13	0.14	177
hb-P15	0.42	55

[a] Thickness of solid film
[b] SHG coefficient.

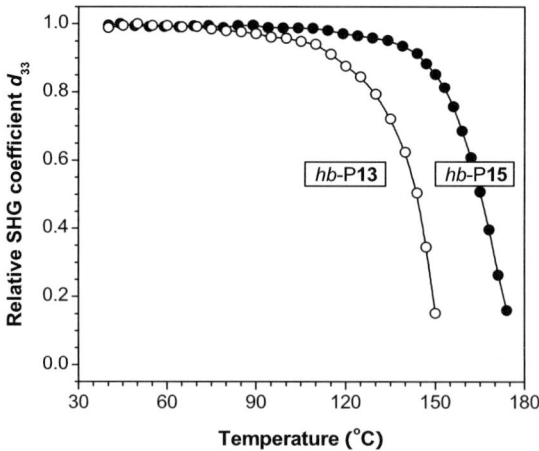

Fig. 8 Decays of SHG coefficients of *hb*-PAEs as a function of temperature

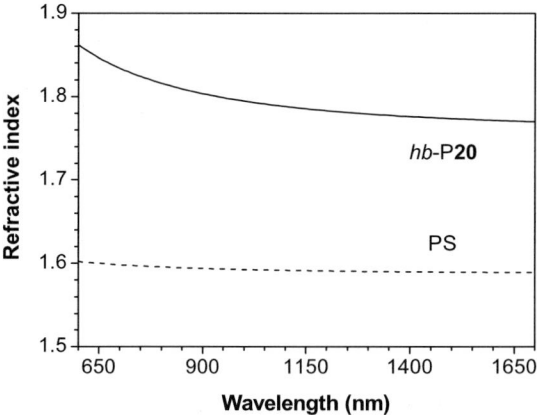

Fig. 9 Wavelength dependence of refractive index of a thin film of *hb*-P20. Data for a thin film of PS is shown for comparison

polarizabilities and small volumes can exhibit high refractivities. Polyyne *hb*-P20 is comprised of electronically mobile aromatic rings and dimensionally slim triple-bond bars and is thus likely to possess high RIs. This proved to the case: a thin film of *hb*-P20 showed RI values of 1.861–1.770 in the spectral region of 600–1700 nm (Fig. 9), which were much higher than those of well-known "organic glasses" such as PS (n = 1.602–1.589), PMMA (n = 1.497–1.489), and PC (n = 1.593–1.576) [131]. The polyyne film was optically transparent and showed high transmittance in the long wavelength region.

3.2
Patterning and Assembling Behaviors

Thin films of the *hb*-PAPs were highly transparent and absorbed almost no visible light. For example, a film of *hb*-P32(5) displayed an optical dispersion as low as 0.009 in the visible region, much superior to those of commercially important organic glasses such as PMMA (0.0175) and polycarbonates (0.0297) [132]. The good film-forming ability and high optical transparency prompted us to utilize the *hb*-PAPs as optical coating materials. The *hb*-PAPs contain many benzyl units, which can readily form radical species, whose recombination will cure or harden the polymers. Indeed, thin films of the hyperbranched polymers were readily cross-linked upon illumination with a UV lamp. Figure 10 shows the formation of insoluble gel upon exposure of a thin film ($\sim$ 1 µm in thickness) of *hb*-P32(4) to a UV irradiation. After $\sim$ 20 min exposure, almost the whole film was cross-linked with an F_g of $\sim$ 100%, indicative of a high photosensitivity despite of its irregular hyperbranched structure.

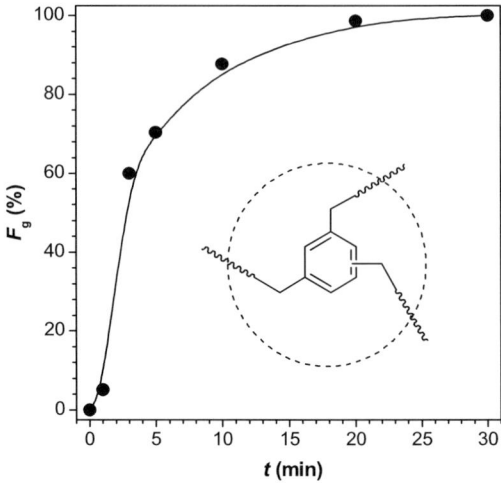

Fig. 10 Plot of gel fraction (F_g) in exposed hb-P32(4) film versus exposure time (t)

Because of its well-known high photoreactivity, benzophenone has been introduced into natural (e.g., protein) and synthetic polymers (e.g., polyimide) to serve as a photo-cross-linker, and its photoreactions in various polymer matrixes such as PS, PMMA, PC and poly(vinyl alcohol) have been well documented [133–136]. As a matter of fact, benzophenone-containing polyimides such as Ciba-Geigy 412 are commercially available high-performance photoresist materials. The hb-PAAs contain numerous aroylbenzene units and, as expected, exhibited very high photo-cross-linking efficiencies. For example, a thin film of hb-P76 on a glass plate could be readily cross-linked by the irradiation with a hand-held UV lamp at room temperature. The cross-linking may have proceeded through the well-established radical mechanism [133–136]: a carbonyl group abstracts a hydrogen atom from a benzyl unit, creating a stable benzyl radical. Coupling or combination of two radicals leads to cross-linking and hence gel formation [105].

Figure 11 depicts the dose effect on the gel formation of hb-PAA films ($l_F = 1–2\,\mu m$) after they have been exposed to a weak UV light with a power of about $1\,mW/cm^2$. Although the photo-cross-linking conditions had not been optimized, all the four hb-PAAs already exhibited much higher sensitivities ($D_{0.5} = 43–180\,mJ/cm^2$) than those of commercial poly(amic ester)-based photoresists ($D_{0.5} = 650–700\,mJ/cm^2$) [137].

Well-resolved patterns with line widths of $\sim 1.0\,\mu m$ were readily formed when a film of hb-P76 was exposed to a UV dose of $1\,J/cm^2$ (Fig. 12A). Patterns with submicron resolutions (line width down to 500 nm) were also achievable, as demonstrated by the examples given in panels B and C of Fig. 12. Clearly, hb-P76 is an excellent photoresist material. Similar to hb-PAAs, hb-PYs were also photosensitive. Well-resolved, defect-free pho-

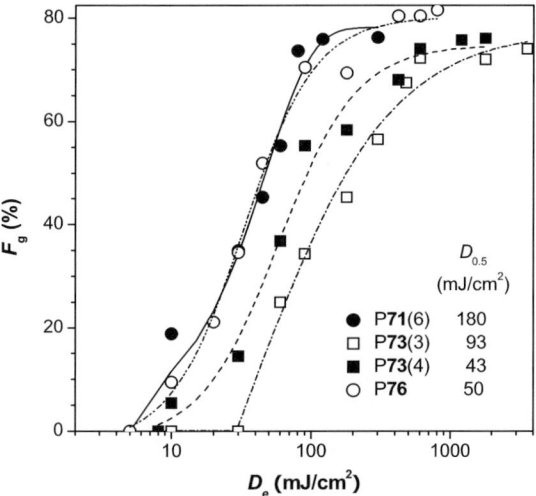

Fig. 11 Plots of gel fractions (F_g) of *hb*-PAA films versus exposure doses (D_e)

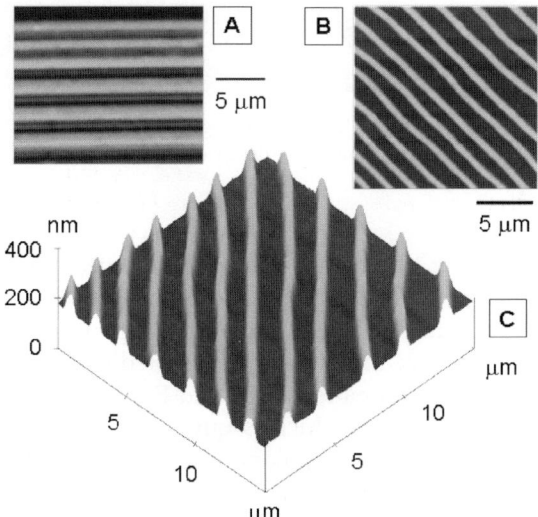

Fig. 12 AFM images of the **A** micro- and **B** and **C** nano-scale patterns obtained from the thin films of *hb*-P76 exposed to 1 J/cm² of UV irradiation

topatterns were readily generated over a large area when a film of *hb*-P20 was exposed to a UV irradiation (Fig. 13, panels A and C). The magnified photograph clearly revealed sharp edges with excellent shape retention.

Conjugated *hb*-PYs constructed from the building blocks containing TPA units were found to show strong PL in solutions upon excitation [34]. Although the photoinduced cross-linking of the diyne units of *hb*-P20 will alter

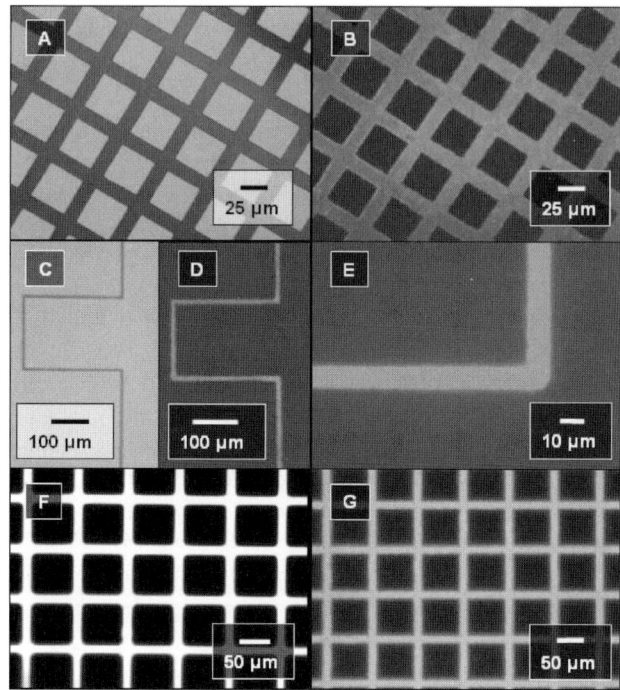

Fig. 13 Negative patterns generated by photolithography of **A–E** hb-P20, **F** hb-r-P[30(4)-20], and **G** hb-1,5-P[30(4)-20]; photographs taken on optical (**A** and **C**) and fluorescence microscopes (**B, D, E** and **G**)

its conjugation structure, the newly formed cumulene, vinylene and ethynylene may maintain or even enlarge the conjugation extents of the resultant polymer gels. We examined the photogenerated patterns of hb-P20 under a fluorescence microscope. The photographs of different cross-linked patterns of hb-P20 are shown in panels B, D, and E of Fig. 13. All the patterned structures emitted bright light, which is quite remarkable, considering that conjugated polymers often show weak emissions in the solid states or when fabricated into thin films due to the non-radiative energy transfer caused by π-stacking of the polymer chains or defect formation [138]. Since the "conventional" photoresists such as SU-8 are generally nonluminescent and can thus only be used as a passive material [139], the bright emission of our polymers may allow them to be used as an active matrix for the fabrications of liquid crystal displays, light-emitting diodes and other photonic devices [140–142]. It is noteworthy that the negative-tone pattern generated by the regiorandom hb-r-PTA containing the TPA core, i.e., hb-r-P[**30**(4)–**20**], emitted a white light (Fig. 13F), whereas its regioregular congener, hb-1,5-P[**30**(4)–**20**], gives blue fluorescence (Fig. 13G). Clearly, the alteration of the substitution pattern has resulted in a change in the effective conjugation length, which offers a new way for fine tuning photonic and electro-optical properties of conjugated polymers through a molecular engineering approach.

Template methods have been widely used to fabricate three-dimensional nano- and microstructured patterns. The breath figure process is an elegant yet simple way to generate large arrays of patterned assemblies [143]. When moist air is blown over a polymer solution in an organic solvent, evaporative cooling generates water droplets on the liquid surface. The uniform droplets arrange into a hexagonal array and sink into the polymer solution. Removal of the volatiles (solvent and water) leaves an imprint of the water droplets as a hollow, air-filled, hexagonally ordered, polymeric bubble array. It has been reported that star-shaped polymers and block copolymers form honeycomb morphologies through the evaporation-induced assembly process [144–146]. Other researchers including us have shown that neither star-shaped nor block structure is necessarily needed for the formation of well-defined assembling morphologies [147–151].

We tried to generate assembling structures of *hb*-PYs by employing the breath figure process. Figure 14 shows the photographs of the patterned structures of *hb*-P20 formed by blowing a stream of moist air over its CS_2 solutions. Hexagonally ordered hollow bubble arrays with void sizes of ca. 10 μm were obtained over a large area. Similar to the patterns generated by UV irradiation through a copper negative mask, the honeycomb patterns obtained from the breath figure process were light-emitting when observed under a fluorescence microscope and can thus potentially be used as an active layer in optical and photonic devices.

As described above, the fabrication of micro- and nano-sized patterns from the hyperbranched polymers as thin layers on defined matrix surfaces has been nicely accomplished. We went one step further and tried to generate free-standing three-dimensional structures. Since *hb*-PAAs can be readily

Fig. 14 Optical micrographs of breath figures of *hb*-P20 obtained from its CS_2 solutions by blow drying in a stream of moist air

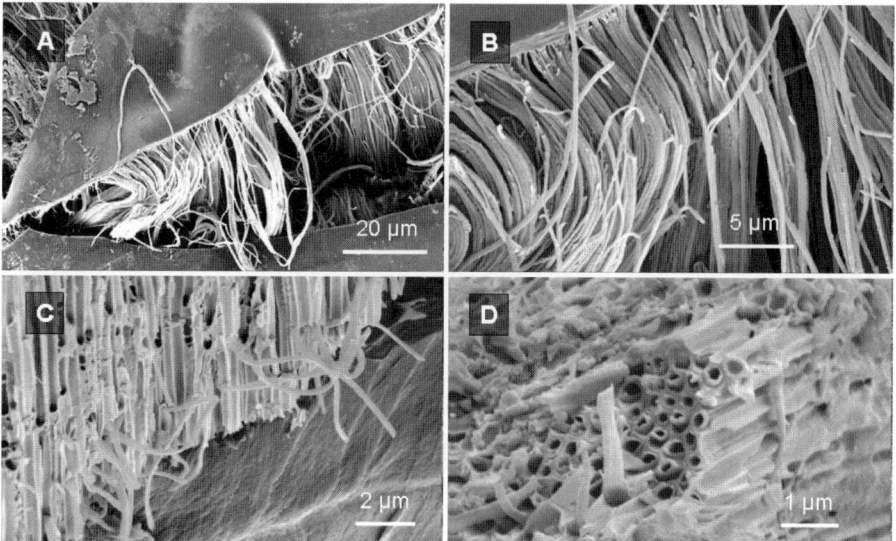

Fig. 15 SEM micrographs of the nanotubes of hb-P71(6) prepared inside an AAO template with a pore size of ~ 250 nm

prepared from the amine-catalyzed polycyclotrimerizations by simply heating the monomer mixtures, we carried out the polymerization of 71(6) in the presence of an AAO template. The great advantage of this system is that it does not suffer from any metal residues left behind after the polymerization and the amine catalyst as well as the solvent can be easily removed. Breakage or dissolution of the AAO templates in aqueous sodium hydroxide solution freed the micro structures. SEM micrographs of the templated hb-P71(6) are shown in Fig. 15. As can be seen from the photos, the hb-PAA adopted well the shape of the template pores and formed micrometer-long polymer nanotubes. The hollows of the nanotubes were clearly confirmed by the magnified images given in Fig. 15D.

3.3
Photonic, Magnetic and Catalytic Properties

Acetylenic molecules are versatile ligands in organometallic chemistry [152–154]. Examples of acetylene-metal reactions include facile complexations of one triple bond with $Co_2(CO)_8$ [155–157] and two triple bonds with $CpCo(CO)_2$ (Scheme 31) [158, 159]. The hb-PYs contain numerous triple bonds and should be readily metallized through their complexations with the cobalt carbonyls. Upon admixing hb-P20 and the cobalt carbonyls in THF at room temperature, the solution color changed from yellow to brown, accompanied by CO gas evolution. The mixtures remained homogenous towards

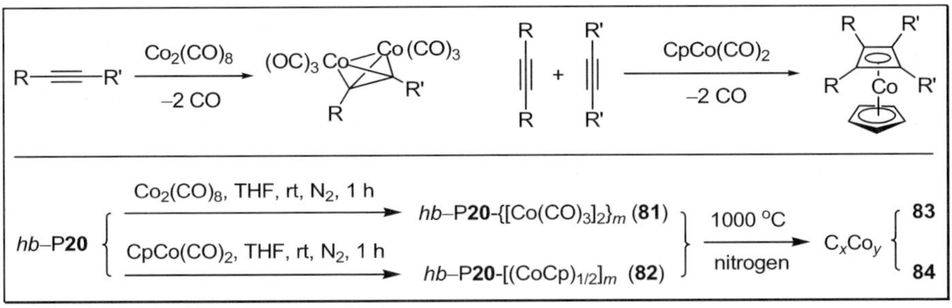

Scheme 31 *Upper panel*: Complex formation between acetylenes and cobalt carbonyls. *Lower panel*: Formation of polymer complexes **81** and **82** via metal complexation and transformation of the complexes into soft ferromagnetic materials **83** and **84** by pyrolytic ceramization

the end of reaction, and the products were purified by pouring the THF solutions into hexane. The polymer complexes are stable in air, whose metal incorporation had been verified by IR analysis [34, 160].

Homogenous yellowish brown-colored films of metallized polymer **81** could be readily prepared by spin coating its freshly prepared complex solution. The films were irradiated through a copper negative mask. Interestingly, the color of the illuminated parts was bleached due to the decomposition of cobalt carbonyl complex, leaving behind the two-dimensional pattern of the photomask (Fig. 16). The enlarged micrograph clearly reveals the sharp edges of the patterns.

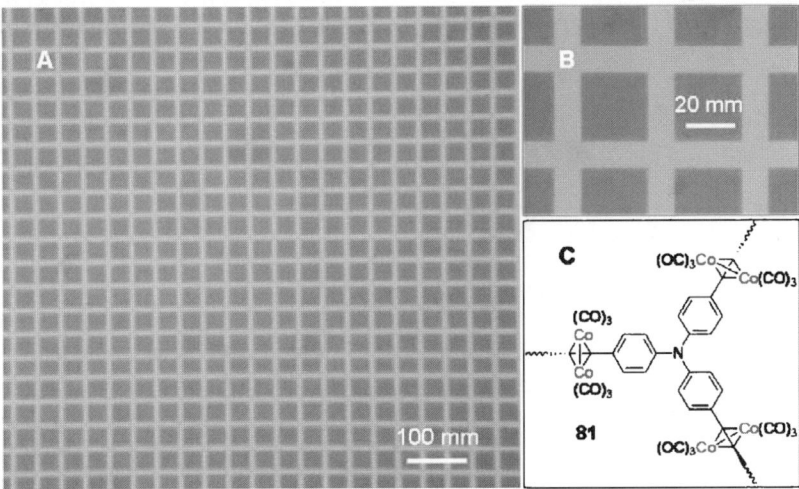

Fig. 16 A Optical micrograph of two-dimensional photopattern generated by photolysis of a metallized *hb*-PY (**81**) through a copper-negative mask. **B** Image with a higher magnification and **C** molecular structure of polymer complex **81**

Inspired by the UV light-induced color change of **81**, we studied its optical properties in more details. Figure 17 shows the wavelength-dependent RI values of its unexposed and exposed films. Similar to its parent *hb*-P20, the metallized polymer exhibits very high RI values ($n = 1.813-1.714$) in the spectral region of 600–1600 nm. Remarkably, the RIs drop significantly after the UV irradiation ($n = 1.777-1.667$). A polymer with such a big RI change is promising for photonic applications: for example, it may serve as photorefractive material in holographic devices [161] or work as high RI optical coating [162].

It has become clear that the carbon-rich *hb*-PYs are readily curable (from ca. 150 °C), thermally stable (up to ca. 550 °C), and pyrolytically carbonizable (yield up to 80% at 900 °C). Furthermore, their triple bonds are easily metallizable by the complexations with cobalt carbonyls. Since the polymer complexes contain a large number of metal atoms, we tried to utilize them as precursors for fabrication of metalloceramics. The pyrolyses of the polyyne-cobalt complexes at 1000 °C for 1 h under nitrogen furnished ceramic products **83** and **84** in 50%–65% yields (cf., Scheme 31). All the ceramics were magnetizable and could be readily attracted to a bar magnet.

The magnetization curves of the magnetoceramics are shown in Fig. 18. With an increase in the strength of the externally applied magnetic field, the magnetization of **83** swiftly increased and eventually leveled off at a saturation magnetization of ~ 118 emu/g, which is much higher than that of the magnet (γ-Fe_2O_3) used in our daily life ($M_s = 74$ emu/g) [163]. The high M_s value of **83**, along with its powder XRD, XPS and SEM data [160], suggests that the cobalt nanocrystallites in the ceramic are well wrapped by carbonaceous species, which have prevented the cobalt nanoparticles from being oxidized during and after the pyrolysis processes [164]. Evidently, the hyperbranched

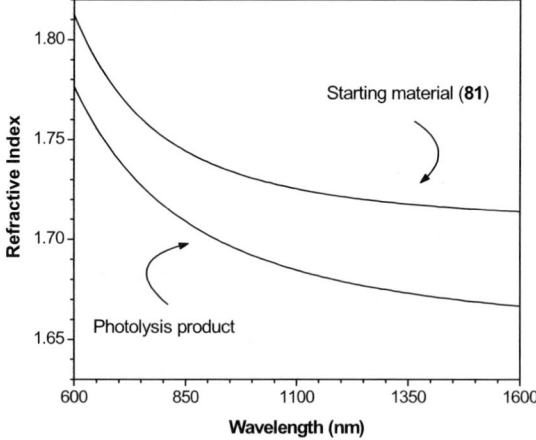

Fig. 17 Refractive indexes of thin films of **81** and its photolysis product

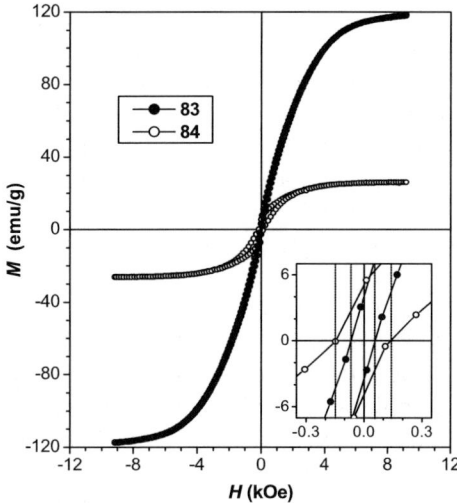

Fig. 18 Plots of magnetization (M) vs. applied magnetic field (H) at 300 K for ceramics **83** and **84**. *Inset*: enlarged portions of the M–H plots in the low strength region of the applied field

polymer complexes are excellent precursors to the magnetic ceramics because their three-dimensional spherical cages have enabled the good retention of the pyrolyzed species and the steady growth of the magnetic crystallites [165]. The M_s value of **84** was somewhat lower ($\sim$ 26 emu/g), which is understandable, because the cobalt content of its precursor complex (**82**) was lower.

Hysteresis loops of the magnetoceramics were very small. From the enlarged H–M plots shown in the inset of Fig. 18, the H_c values of **83** and **84** were found to be 0.058 and 0.142 kOe, respectively. The high magnetizability (M_s up to 118 emu/g) and low coercivity (H_c down to $\sim$ 0.06 kOe) of **83** make it an excellent soft ferromagnetic material, which may have an array of high-technology applications in various electromagnetic systems.

From our previous studies, we knew that hyperbranched polymers containing ferrocene units can be transformed into nano-sized magnetoceramics [165–167]. Taking the advantages of the thermal stability and cross-linking capability of the ferrocene-containing *hb*-PAAs, we first generated organometallic micropatterns by developing a thin film of *hb*-P(**76**-XIII) in dichloroethane, after the film had been exposed to a UV irradiation through a copper negative mask for 30 min. In the second step, these patterns were readily transformed into magnetoceramics by pyrolyzing the microscale grid structures at 1000 °C for 1 h under nitrogen. As can be seen from Fig. 19, the negative-tone photoresist was readily transformed to well-defined, microscopically patterned, iron-containing ceramics with excellent shape retention. Similar to their counterparts prepared in the bulk, the patterned ceramic was also soft ferromagnetic.

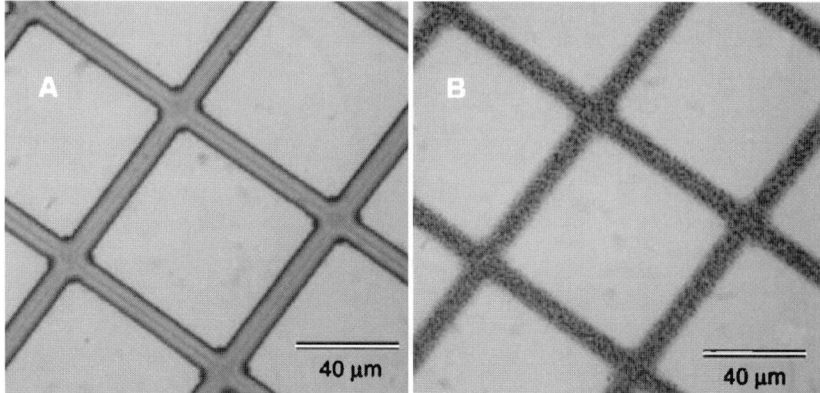

Fig. 19 Photographs of **A** the micropattern fabricated by the photolysis of ferrocene-containing *hb*-P(76-XIII) using a copper negative mask and **B** the magnetoceramic pattern generated by the pyrolysis of the micropattern under nitrogen at 1000 °C

Metallic species such as iron, nickel and cobalt are known to catalyze the growth of CNTs in the CVD process [168–170]. Because of the ready thermal curability of the *hb*-PYs, spin-coated films of organometallic polymers **81** and **82** are expected to restrict the agglomeration of the metallic nanoclusters in the CVD process and hence to provide nanoscopic catalyst seeds for the

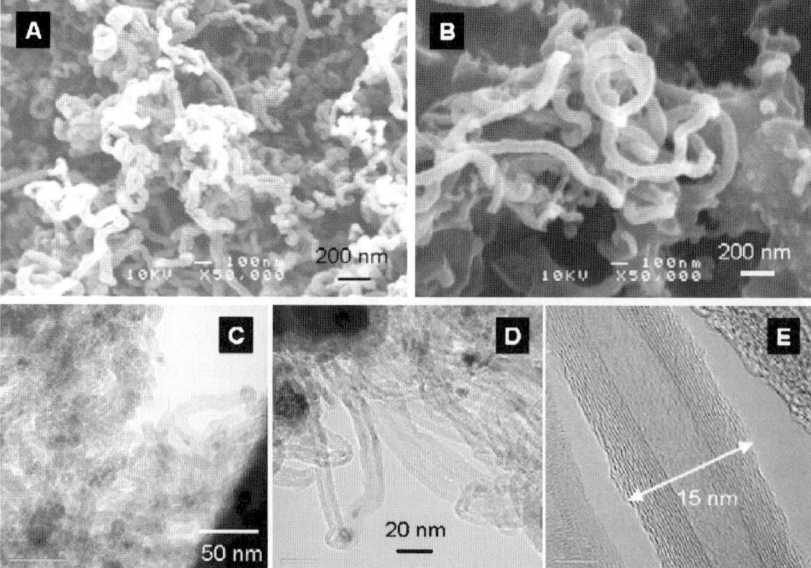

Fig. 20 **A** and **B** SEM and **C–E** TEM micrographs of the CNTs prepared by a CVD process on the silicon wafers spin-coated with **81**

CNT growth. This proved to be the case. Uniform bundles of CNTs were grown by the CVD process at 700 °C with acetylene gas as the carbon source (Fig. 20). The diameters and lengths of the CNTs were alterable by variation of the surface activation as well as the growth time [171]. As expected from a CNT growing temperature below 800 °C [172, 173], the formed CNTs were multiwalled in nature. Their diameters were, however, as small as 15 nm (Fig. 20E).

4
Concluding Remarks

In this review we have briefly summarized the results on the synthesis of a variety of functional hyperbranched polymers from acetylene triple-bond building blocks. Effective polymerization reactions including polycoupling, polyhydrosilylation, polycycloaddition and polycyclotrimerization of mono-, di- and triynes catalyzed by metallic and non-metallic species have been developed, which have enabled the creation of hyperbranched polyynes, polyenes, polyarylenes and polytriazoles with high molecular weights and macroscopic processability in high yields. The new polymerization routes opened and the new structural insights gained offer versatile synthetic tools and valuable guidelines for further developments in this area of research.

Using the triple-bond building blocks in combination with functional groups have resulted in the generation of conjugated polymers with functional properties. The carbon-rich polymers exhibited outstanding thermal stabilities. Incorporation of chromophoric units enabled the modulation of emission colors and efficiencies of the polymers at the molecular level. The numerous aromatic rings in the *hb*-PAs conferred high optical limiting power on the polymers, while the poled films of the azo-functionalized *hb*-PAEs showed stable NLO performance with high SHG coefficients. Combination of the slim diyne linker with the polarizable TPA core results in high optical transparency with exceptionally high photorefractivity. The photosensitive benzyl, benzophenone and diyne units endowed the polymers with photocurability and hence the potential to be used as photoresists and active matrixes in optical devices. The assemblies of hexagonal arrays of breath figures and the micrometer-long polymer nanotubes were obtained from the dynamic and static templating processes. The complexation with the cobalt carbonyls yielded spin-coatable hyperbranched organometallic polymers, whose RIs were readily tuned by UV irradiation. The hyperbranched polymer complexes also served as precursors to advanced soft ferromagnetic ceramics and as catalysts for the growth of CNTs.

Their simple syntheses and ready processability, coupled with their unique structures and useful functionalities, make this group of hyperbranched polymers attractive and promising for an array of high-technology applications.

Acknowledgements This work was partially supported by the Research Grants Council of Hong Kong, the National Science Foundation of China, and the Ministry of Science and Technology of China. We thank all the people involved in this project, some of whose names are given in the references. B.Z.T. thanks the support from the Cao Guangbiao Foundation of the Zhejiang University.

References

1. Flory PJ (1952) J Am Chem Soc 74:2718
2. Wang F, Wilson MS, Rauh RD, Schottland P, Reynolds JR (1999) Macromolecules 33:4272
3. Gao C, Yan D (2001) Macromolecules 34:156
4. Emrick T, Chang HT, Frechet JMJ (1999) Macromolecules 32:6380
5. Jikei M, Chon SH, Kakimoto M, Kawauchi S, Imase T, Watanabe J (1999) Macromolecules 32:2061
6. Russo S, Boulares A, da Rin A, Mariani A, Cosulich ME (1999) Macromol Symp 143:309
7. Ahoni SM (1995) Polym Adv Technol 6:373
8. Frechet JMJ, Henmi M, Gitsov I, Aoshima S, Leduc M, Grubbs RB (1995) Science 269:1080
9. Hawker CJ, Frechet JMJ, Grubbs RB, Dao J (1995) J Am Chem Soc 117:10763
10. Gaynor SG, Edelman SZ, Matyjaszewski K (1996) Macromolecules 29:1079
11. Dworak A, Walach W, Trzebicka B (1995) Macromol Chem Phys 196:1963
12. Suzuki M, Yoshida S, Shiraga K, Saegusa T (1998) Macromolecules 31:1716
13. Magnusson H, Malmström E, Hult A (1999) Macromol Rapid Commun 20:453
14. Bednarek M, Biedron T, Helinski J, Kaluzynski K, Kubisa P, Penczek S (1999) Macromol Rapid Commun 20:369
15. Sunder A, Hanselmann R, Frey H, Mülhaupt R (1999) Macromolecules 32:4240
16. Sunder A, Heinemann J, Frey H (2000) Chem Eur J 6:2499
17. Hult A, Johansson M, Malström E (1999) Adv Polym Sci 143:1
18. Voit B (2003) Comptes Rendus Chimie 6:821
19. Bunz UHF (2001) Acc Chem Res 34:998
20. Negishi EI, Anastasia L (2003) Chem Rev 103:1979
21. Bharathi P, Moore JS (1997) J Am Chem Soc 119:3391
22. Bharathi P, Moore JS (2000) Macromolecules 33:3212
23. Kim C, Chang Y, Kim JS (1996) Macromolecules 29:6353
24. Fomina L, Salcedo R (1996) Polymer 37:1723
25. Hittinger E, Kokil A, Weder C (2004) Angew Chem Int Ed 43:1808
26. Weder C (2005) Chem Commun, p 5378
27. Dong YQ, Li Z, Lam JWY, Dong YP, Feng XD, Tang BZ (2005) Chinese J Polym Sci 23:665
28. Li Z, Qin A, Lam JWY, Dong YQ, Dong YP, Ye C, Williams ID, Tang BZ (2006) Macromolecules 39:1436
29. Jikei M, Chon SH, Kakimoto M, Kawauchi S, Imase T, Watanabe J (1999) Macromolecules 32:2061
30. Lin Q, Long TE (2003) Macromolecules 36:9809
31. Bunz UHF (2000) Chem Rev 100:1605
32. Onitsuka K, Ohshiro N, Fujimoto M, Takei F, Takahashi S (2000) Mol Cryst Liq Cryst 342:159

33. Khan MS, Schwartz DJ, Pasha NA, Kakkar AK, Lin B, Raithby PR, Lewis JZ (1992) Anorg Allg Chem 616:121
34. Häussler M, Zheng R, Lam JWY, Tong H, Dong H, Tang BZ (2004) J Phys Chem B 108:10645
35. Ogawa T (1995) Prog Polym Sci 20:943
36. Hay AS (1998) J Polym Sci Part A: Polym Chem 36:505
37. Hoffmann B, Zanini D, Ripoche I, Burli R, Vasella A (2001) Helv Chim Acta 84:1862
38. Siemsen P, Livingston RC, Diederich F (2000) Angew Chem Int Ed 39:2632
39. Klebansky AL, Grachev IV, Kuznetsova OM (1957) J Gen Chem USSR 27:3008
40. Bohlmann F, Schönowsky H, Inhoffen E, Grau G (1964) Chem Ber 97:794
41. Hawker CJ, Lee R, Frechét JMJ (1991) J Am Chem Soc 113:4583
42. Hölter D, Burgath A, Frey H (1997) Acta Polym 48:30
43. Xiao Y, Wong RA, Son DY (2000) Macromolecules 33:7232
44. Kwak G, Masuda T (2002) Macromol Rap Commun 23:68
45. Kwak G, Takagi A, Fujiki M, Masuda T (2004) Chem Mater 16:781
46. Rao TV, Yamashita H, Uchimaru Y, Sugiyama J, Takeuchi K (2005) Polymer 46:9736
47. Morgenroth F, Müllen K (1997) Tetrahedron 53:15349
48. Huisgen R (1984) In: Padwa A (ed) 1,3-Dipolar Cycloadditions: Introduction, Survey, Mechanism. Wiley, New York
49. Huisgen R, Szeimies G, Moebius L (1967) Chem Ber 100:2494
50. Kolb HC, Finn MG, Sharpless KB (2001) Angew Chem Int Ed 40:2004
51. Rostovtsev VV, Green LG, Fokin VV, Sharpless KB (2002) Angew Chem Int Ed 41:2596
52. Scheel A, Komber H, Voit B (2004) Macromol Rapid Commun 25:1175
53. Voit B (2005) J Polym Sci Part A: Polym Chem 43:2679
54. Qin A, Haeussler M, Lam JWY, Tse KKC, Tang BZ (2006) Polym Prepr 47(2):681
55. Chan TR, Hilgraf R, Sharpless KB, Fokin VV (2004) Org Lett 6:2853
56. Himo F, Lovell T, Hilgraf R, Rostovtsev VV, Noodleman L, Sharpless KB, Fokin VV (2005) J Am Chem Soc 127:210
57. Zhang L, Chen X, Xue P, Sun HHY, Williams ID, Sharpless KB, Fokin VV, Jia G (2005) J Am Chem Soc 127:15998
58. Majireck MM, Weinreb SM (2006) J Org Chem 71:8680
59. Sergeyev VA, Shitikov VK, Chernomordik YA, Korshak VV (1975) Appl Polym Symp 26:237
60. Sergeyev VA, Shitikov VK, Kurapov AS, Antonova-Antipova IP (1989) Polym Sci USSR 31:1300
61. Srinrivasan R, Farona MF (1988) Polym Bull 20:359
62. Chalk AJ, Gilbert AR (1972) J Polym Sci: Part A-1 10:2033
63. Bracke W (1972) J Polym Sci: Part A-1 10:2097
64. Xu K, Tang BZ (1999) Chinese J Polym Sci 17:397
65. Xu K, Peng H, Sun Q, Dong Y, Salhi F, Luo J, Chen J, Hunag Y, Zhang D, Xu Z, Tang BZ (2000) Macromolecules 35:5821
66. Zheng R, Dong H, Peng H, Lam JWY, Tang BZ (2004) Macromolecules 37:5196
67. Xu K, Peng H, Huang Y, Xu Z, Tang BZ (2000) Polym Prepr 41(2):1245
68. Li Z, Lam JWY, Dong YQ, Dong YP, Sung HHY, Williams ID, Tang BZ (2006) Macromolecules 39:6458
69. Saito S, Yamamoto Y (2000) Chem Rev 100:2901
70. Lautens M, Klute W, Tam W (1996) Chem Rev 96:49
71. Melikyan GG, Nicholas KM (1995) In: Stang PJ, Diederich F (eds) Modern Acetylene Chemistry. Wiley, Weinheim, p 99

72. Schore NE (1991) In: Trost BM, Flemming I (eds) Comprehensive Organic Synthesis. Pergamon, Oxford 5:1129
73. Vollhardt KPC (1984) Angew Chem Int Ed 23:539
74. Mori N, Ikeda S, Odashima K (2001) Chem Commun p 181
75. Tekeuchi R, Tanaka S, Nakaya Y (2001) Tetrahedron Lett 42:2991
76. Yamamoto Y, Ogawa R, Itoh K (2000) Chem Commun, p 549
77. Ozerov OV, Ladipo FT, Patrick BO (1999) J Am Chem Soc 121:7941
78. Johnson ES, Balaich GJ, Rothwell IP (1997) J Am Chem Soc 119:7685
79. Masuda T, Mouri T, Higashimura T (1980) Bull Chem Soc Jpn 53:1152
80. Cotton FA, Hall WT (1980) Inorg Chem 19:2352
81. Bruck MA, Copenhaver AS, Wigley DE (1987) J Am Chem Soc 109:6525
82. Cotton FA, Hall WT (1981) Inorg Chem 20:1285
83. Strickler JR, Wexler PA, Wigley DE (1988) Organometallics 7:2067
84. Cotton FA, Wilkinson G (1988) Advanced Inorganic Chemistry, 5th ed, Chap 19B. Wiley, New York
85. Jhingan AK, Maier WF (1987) J Org Chem 52:1161
86. Zheng R, Häussler M, Dong H, Lam JWY, Tang BZ (2006) Macromolecules 39:7973
87. Häußler M, Lam JWY, Peng H, Zheng R, Tang BZ (2003) Polym Prepr 44(1):1177
88. Häußler M, Lam JWY, Zheng R, Peng H, Luo J, Chen J, Law CCW, Tang BZ (2003) Comptes Rendus Chimie 6:833
89. Law CCW, Chen J, Lam JWY, Peng H, Tang BZ (2004) J Inorg Organomet Polym 14:39
90. Chen J, Peng H, Law CCW, Dong YP, Lam JWY, Williams ID, Tang BZ (2003) Macromolecules 36:4319
91. Zheng R, Dong H, Tang BZ (2005) In: Abd-El-Azi A, Carraher C, Pittman C, Sheats J, Zeldin M (eds) Macromolecules Containing Metal- and Metal-like Elements. Wiley, New York
92. Peng H, Luo J, Cheng L, Lam JWY, Xu K, Dong Y, Zhang D, Huang Y, Xu Z, Tang BZ (2002) Opt Mater 21:315
93. Peng H, Cheng L, Luo J, Xu K, Sun Q, Dong Y, Salhi F, Lee PPS, Chen J, Tang BZ (2002) Macromolecules 35:5349
94. Peng H, Lam JWY, Tang BZ (2005) Polymer 46:5746
95. Peng H, Lam JWY, Tang BZ (2005) Macromol Rapid Commun 26:673
96. Peng H, Zheng R, Dong H, Jia D, Tang BZ (2005) Chin J Polym Sci 23:1
97. Häußler M, Dong H, Lam JWY, Zheng R, Qin A, Tang BZ (2005) Chin J Polym Sci 23:567
98. Dong H, Lam JWY, Häußler M, Zheng R, Peng H, Law CCW, Tang BZ (2004) Curr Trends Polym Sci 9:15
99. Kong X, Lam JWY, Tang BZ (1999) Macromolecules 32:1722
100. Zheng R, Lam JWY, Peng H, Häußler M, Law CCW, Tang BZ (2003) Polym Prepr 44(2):770
101. Balasubramanian K, Selvaraj S, Venkataramani PS (1980) Synthesis p 29
102. Pigge FC, Ghasedi F, Rath NJ (2002) J Org Chem 67:4547
103. Matsuda K, Nakamura N, Iwamura H (1994) Chem Lett, p 1765
104. Dong H, Zheng R, Lam JWY, Häußler M, Tang BZ (2004) Polym Prepr 45(2):825
105. Dong H, Zheng R, Lam JWY, Häußler M, Tang BZ (2005) Macromolecules 38:6382
106. Saito S, Yamamoto Y (2000) Chem Rev 100:2901
107. Jikei M, Kakimoto M (2001) Prog Polym Sci 26:1233
108. Schluter AD, Wegner G (1993) Acta Polym 44:59
109. Tour JM (1994) Adv Mater 6:190
110. Johnen NA, Kim HK, Ober CK (1994) ACS Symp Ser 579:298

111. Kumar U, Neenan TX (1995) ACS Symp Ser 614:4084
112. Watson MD, Fechtenkotter A, Mullen K (2001) Chem Rev 101:1267
113. Masuda T, Tang BZ, Higashimura T, Yamaoka H (1985) Macromolecules 18:2369
114. Badarau C, Wang ZY (2004) Macromolecules 37:147
115. Armistead JP, Houser EJ, Keller TM (2000) Appl Organomet Chem 14:253
116. Gandon S, Mison P, Sillion B (1996) ACS Symp Ser 624:306
117. Rutherford DR, Stille JK, Elliott CM, Reichert VR (1992) Macromolecules 25:2294
118. Hergenrother PM (1990) In: Kroschwitz JI (ed) Concise Encyclopedia of Polymer Science & Engineering. Wiley, New York, p 5
119. Beckham HW, Keller TM (2002) J Mater Chem 12:3363
120. Mochizuki H Hasui T, Kawamoto M, Ikeda T, Adachi C, Taniguchi Y, Shirota Y (2003) Macromolecules 36:3457
121. Berlman IB (1971) Handbook of Fluorescence Spectra of Aromatic Molecules, 2nd ed. Academic Press, New York
122. Chen J, Xie Z, Lam JWY, Law CCW, Tang BZ (2003) Macromolecules 36:1108
123. Tang BZ, Leung SM, Peng H, Yu NT, Su KC (1997) Macromolecules 30:2848
124. Lee M, Katz HE, Erben C, Gill DM, Gopalan P, Heber JD, McGee DJ (2002) Science 298:1401
125. Shi Y, Zhang C, Zhang H, Bechtel JH, Dalton LR, Robinson BH, Steier WH (2000) Science 288:119
126. Burland DM, Miller RD, Walsh CA (1994) Chem Rev 94:31
127. Marder SR, Cheng LT, Tiemann BG, Friedli AC, Blanchard-Desce M, Perry JW, Skindhøj J (1994) Science 263:511
128. Dalton LR, Harper AW, Ren A, Wang F, Todorova G, Chen J, Zhang C, Lee M (1999) Ind Eng Chem Res 38:8
129. Luo J, Ma H, Haller M, Barto RR (2002) Chem Commun, p 888
130. Seferis JC (1989) Refractive Indices of Polymers. In: Brandrup J, Immergut EH (eds) Polymer Handbook, 3rd ed. Wiley, New York, p VI/451
131. Mills NJ (1990) In: Kroschwitz JI (ed) Concise Encyclopedia of Polymer Science & Engineering. Wiley, New York, p 683
132. Xu K, Peng H, Tang BZ (2001) Polym Prepr 42(1):555
133. Luo Y, Leszyk J, Qian YD, Gergely J, Tao T (1999) Biochemistry 38:6678
134. Hasegawa M, Horie K (2001) Prog Polym Sci 26:259
135. Turro NJ (1978) Modern Molecular Photochemistry. In: Benjamin-Cummings Publ. Co. (ed) Modern Molecular Photochemistry. Menlo Park, California
136. Qu BJ, Xu YH, Shi FW (1992) Macromolecules 25:5215
137. Kim K-H, Jang S, Harris FW (2001) Macromolecules 34:8925
138. Hua JL, Lam JWY, Dong H, Wu L, Wong KS, Tang BZ (2006) Polymer 47:18
139. Wang Y, Bachman M, Sims CE, Li GP, Allbritton NL (2006) Langmuir 22:2719
140. Kim C, Burrows PE, Forrest SR (2000) Science 288:831
141. Sirringhaus H, Kawase T, Friend RH, Shimoda T, Inbasekaran M, Wu W, Woo EP (2000) Science 290:2123
142. Garnier F, Hajlaoui R, Yassar A, Srivastava P (1994) Science 265:1684
143. Bunz UHF (2006) Adv Mater 18:973
144. Widawski G, Rawiso M, Francois B (1994) Nature 369:387
145. Jenekhe SA, Chen XL (1999) Science 283:372
146. de Boer B, Stalmach U, Nijland H, Hadziioannou G (2000) Adv Mater 12:1581
147. Srinivasarao M, Collings D, Philips A, Patel S (2001) Science 292:79
148. Peng J, Han Y, Li B (2004) Polymer 45:447
149. Tang BZ (2001) Polym News 26:262

150. Lam JWY, Tang BZ (2005) Acc Chem Res 38:745
151. Salhi F, Cheuk KKL, Sun Q, Lam JWY, Cha JAK, Li G, Li B, Luo J, Chen J, Tang BZ (2001) J Nanosci Nanotechnol 1:137
152. Babudri F, Farinola GM, Naso F (2004) J Mater Chem 14:11
153. Long NJ, Williams CK (2003) Angew Chem Int Ed 42:2586
154. Bunz UHF (2003) J Organomet Chem 683:269
155. Newkome GR, He EF, Moorefield CN (1999) Chem Rev 99:1689
156. Chauhan BPS, Corriu RJP, Lanneau GF, Priou C, Auner N, Handwerker H, Herdtweck E (1995) Organometallics 14:1657
157. Chan WY, Berenbaum A, Clendenning SB, Lough AJ, Manners I (2003) Organometallics 22:3796
158. Nishihara H, Kurashina M, Murata M (2003) Macromol Symp 196:27
159. Altmann M, Bunz UHF (1995) Angew Chem Int Ed 34:569
160. Häußler M, Lam JWY, Zheng R, Dong H, Tong H, Tang BZ (2005) J Inorg Organomet Polym Mat 15:519
161. Hendrickx E, Engels C, Schaerlaekens M, Van Steenwinckel D, Samyn C, Persoons A (2002) J Phys Chem B 106:4588
162. Lu C, Guan C, Liu Y, Cheng Y, Yang B (2005) Chem Mater 17:2448
163. Tang BZ, Geng Y, Lam JWY, Li B, Jing X, Wang X, Wang F, Pakhomov AB, Zhang XX (1999) Chem Mater 11:1581
164. O'Handley RC (2000) Modern Magnetic Materials: Principles and Applications. Wiley, New York, p 491
165. Sun Q, Xu K, Peng H, Zheng R, Häußler M, Tang BZ (2003) Macromolecules 36:2309
166. Sun Q, Lam JWY, Xu K, Xu H, Cha JAK, Zhang X, Jing X, Wang F, Tang BZ (2000) Chem Mater 12:2617
167. Häußler M, Sun Q, Xu K, Lam JWY, Dong H Tang BZ (2005) J Inorg Organomet Polym 15:67
168. Chatterjee AK, Sharon M, Baneriee R, Neumann-Spallart M (2003) Electrochim Acta 48:3439
169. Huang ZP, Wang DZ, Wen JG, Sennett M, Gibson H, Ren ZF (2002) Appl Phys A 74:387
170. Deck CP, Vecchio K (2006) Carbon 44:267
171. Häußler M, Tse KC, Lam JWY, Tong H, Qin A, Tang BZ (2006) Polym Mat Sci Eng 95:213
172. Lu JQ, Kopley TE, Moll N, Roitman D, Chamberlin D, Fu Q, Liu J, Russell TP, Rider DA, Manners I, Winnik MA (2005) Chem Mater 17:2227
173. Lu JQ, Rider DA, Onyegam E, Wang H, Winnik MA, Manners I, Cheng Q, Fu Q, Liu J (2006) Langmuir 22:5174

Editor: Kwang-Sup Lee

Polymer Monolayer Dynamics

Alan R. Esker[2] · Chanjoong Kim[3] · Hyuk Yu[1] (✉)

[1] Department of Chemistry, University of Wisconsin, Madison, WI 53706, USA
yu@chem.wisc.edu

[2] *Present address:*
Department of Chemistry, Virginia Polytechnic Institute and State University, Blacksburg, VA 24061, USA

[3] *Present address:*
Department of Physics & Division of Engineering & Applied Sciences, Harvard University, Cambridge, MA 02138, USA

1	Introduction: Brief History of Monolayers	60
2	Static Properties of Polymer Monolayers	61
3	Capillary Wave Dynamics	65
4	Surface Light Scattering Method	75
5	Polymer Systems	80
5.1	Homopolymers	80
5.1.1	Vinyl Polymers and Polyethers	80
5.1.2	Binary Monolayer: Side Chain Length Effect	88
5.1.3	Temperature Dependence	91
5.2	Copolymers	92
5.2.1	Alternating Copolymers	92
5.2.2	Block Copolymers	98
6	Other Methods for Monolayer Dynamics	104
7	Conclusions	106
	References	107

Abstract This is to review viscoelastic properties of monomolecular layers of polymers on the air/water interface, as probed principally by surface light scattering. The method is a non-invasive one that makes use of spontaneous capillary waves, induced by density fluctuations within liquids under thermal equilibrium. The capillary waves are also called ripplons, and they propagate with temporal damping. The interface is determined to be molecularly smooth but still dielectric permittivity difference between air and water is large enough to give rise to strong light scattering. Thus, the scattering amounts to a surface analog of Brillouin scattering in bulk liquid wherein spontaneously propagating phonons interact with light. Thus, the power spectra of scattered light from the interface provide the propagation rate and the damping coefficient. Analysis is based on the resonant mode-coupling of lateral and transverse waves that are recast into the lateral storage modulus and the corresponding loss modulus. By virtue of the two-dimensional character of the monolayers, many intriguing observations have been made with respect to amphiphilic properties and chain architecture of homopolymers and copolymers. Close connection and correspondence between the static properties of polymer monolayers and their rheological behavior have been established, and the review covers reports over the past two decades.

Keywords Monolayers · Surface light scattering · Capillary waves · Dispersion equation · Dilational elastic modulus · Dilational loss modulus · Scaling exponent

Abbreviations
2D	two-dimension
A/W	the air/water interface
CMC	critical micelle concentration
ISR	interface stress rheometer
O/W	oil/water interface
PEO	poly(ethylene oxide)
PDcMA	poly(1-decene-co-maleic acid)
PHcMA	poly(1-hexene-co-maleic acid)
PMA	poly(methyl acrylate)
PMMA	poly(methyl methacrylate)
PnBMA	poly(n-butyl methacrylate)
POcMA	poly(1-decene-co-maleic acid)
PODcMA	poly(1-octadecene-co-maleic acid)
PtBMA	poly(t-butyl methacrylate)
PVAc	poly(vinyl acetate)
PVP	poly(vinyl palmitate)
PVS	poly(vinyl stearate)
PTHF	poly(tetrahydrofuran)
SLS	surface light scattering

1
Introduction: Brief History of Monolayers

This is a time of ferment for science and technology of surfaces and interfaces, in part driven by interest in nanometer scale objects. Attending to this ferment, there exists a driving force for quantitative understanding of polymers at surfaces and interfaces since a significant fraction of formulation of new functional materials and sensors starts with macromolecules of synthetic and biological origins as brushes and mushrooms on various substrates [1]. Monomolecular layers on the air/water interface (A/W) can be regarded as primordial interfacial objects. Hence, this article attempts to review what we have learned for the past two decades about the dynamics of such interfacial objects, namely polymer monolayers on A/W.

The phenomenon of oil on a water surface having calming effects on its ripple has been known since ancient Greece [2]. The first documented observation is attributed to Benjamin Franklin [3] in London who poured a teaspoonful of oil on Clapham Pond and found its surface becoming mirror smooth. The actual event is estimated to have taken place between 1769 and 1771, and the account appeared in 1774 in Philosophical Transaction of the Royal Society [3–6]. Following the account of this experiment, many illustrious names in 19th century science are associated with the phenomena of layering amphiphilic substances on water surfaces such as Lord Rayleigh [7], Kelvin [8], Pockels [9], and others. The intriguing history is presented with a dramatic flair by Giles, Giles and Forrester [4–6] and Tanford with his usual inimitable perspective [2]. Monolayers at A/W of macromolecules of biological and synthetic origins had begun to be studied in the middle of the last century. The first examinations of polymer monolayers are traced to Crisp [10, 11] in 1946 in J Colloid Sci and subsequently to Gabrielli and Pugelli in 1971 [12] when they compared how the area per monomer unit changes from A/W to an oil/water interface. Resurgence of interest in polymer monolayers, particularly its dynamics in recent decades has technological underpinnings. The interest is in part derived from a search for optimum conditions for transfer of monolayers to solid substrates to form Langmuir–Blodgett films [13–15] in various non-linear optical applications [16, 17]. The technological potential has diminished mainly due to the life-time problems of devices, while fundamental questions remain relative to the rheology of monolayers as two-dimensional objects.

2
Static Properties of Polymer Monolayers

In order to set the stage for this review of the polymer dynamics on monolayers at interfaces with emphasis on A/W, we need to lay out its static properties first. Surface tension σ represents a fundamental property of a liquid surface. The change in Gibbs free energy dG for a multi-component system including the surface contribution is written as

$$dG = -S\,dT + V\,dp + \sigma\,dA_s + \sum_i \mu_i\,dn_i, \qquad (1)$$

where S is the total entropy of the system, T temperature, V volume, p pressure, A_s surface area, and μ_i and n_i correspond, respectively, to the chemical potential and number of moles of the ith component. Thus, the surface tension is defined as the incremental change in the Gibbs free energy with respect to the surface area change,

$$\sigma = \left(\frac{\partial G}{\partial A_S}\right)_{T,p} \qquad (2)$$

The presence of a surface film or monolayer on a pure liquid is expressed by the surface pressure Π, that is defined as

$$\Pi \equiv \sigma_{\text{pure liquid surface}} - \sigma_{\text{film covered surface}} \,. \tag{3}$$

It should be pointed out at this juncture that strict thermodynamics treatment of the film-covered surfaces is not possible [18]. The reason is difficulty in delineation of the system. The interface, typically of the order of a 1–2 nm thick monolayer, contains a certain amount of bound water, which is in dynamic equilibrium with the bulk water in the subphase. In a strict thermodynamic treatment, such an interface must be accounted as an open system in equilibrium with the subphase components, principally water. On the other hand, a useful conceptual framework is to regard the interface as a "2-dimensional" (2D) object such as a 2D gas or 2D solution [19, 20]. Thus, the surface pressure Π is treated as either a 2D gas pressure or a 2D osmotic pressure. With such a perspective, an analog of either p-V isotherm of a gas or the osmotic pressure-concentration isotherm, Π-c, of a solution is adopted. It is commonly referred to as the surface pressure-area isotherm, Π-A, where A is defined as an average area per molecule on the interface, under the provision that all molecules reside in the interface without desorption into the subphase or vaporization into the air. A more direct analog of Π-c of a bulk solution is Π - Γ where Γ is the mass per unit area, hence is the reciprocal of A, the area per unit mass. The nature of the collapsed state depends on the solubility of the surfactant. For truly insoluble films, the film collapses by forming multilayers in the upper phase. A broad illustrative sketch of a Π-Γ plot is given in Fig. 1.

For most monolayers, the range of surface pressures for which gas model analogs are applicable, is quite low ($\Pi < 1$ mN m^{-1}). The simplest model one could consider is an analog to an ideal gas or an ideal solution,

$$\Pi A = kT \,, \tag{4}$$

or

$$\Pi = \frac{\Gamma RT}{M_n} \tag{5}$$

that is the analog of the ideal gas equation of state, $pv = kT$, with v as the molecular volume and k the Boltzmann constant, or that of ideal solution, $\Pi = cRT/M_n$, with c as the mass concentration and R the gas constant. Expanding on the ideal solution analog into a virial form, we can add the second virial coefficients A_2, such that

$$\Pi = \frac{RT\Gamma}{M_n} \left(1 + A_2 \Gamma + ...\right) \,. \tag{6}$$

Such an expansion was employed by many in the early 1980s [21–27]. It is, however, of little routine use for the static characterization of polymeric monolayers since accurate measurements of surface pressure in this surface

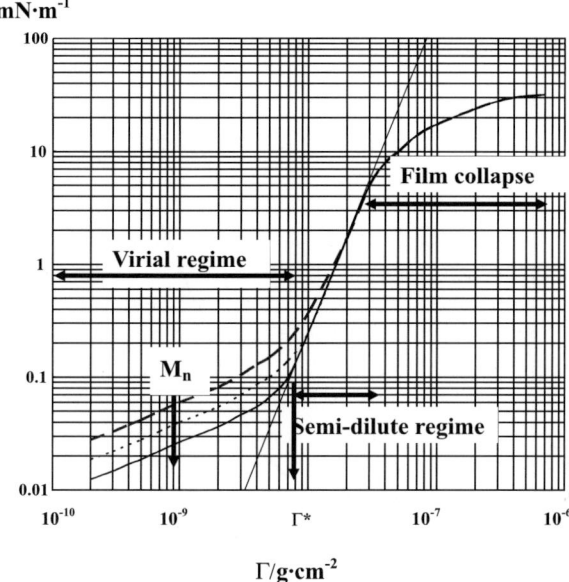

Fig. 1 Generic isotherms for polymeric monolayers. Surface pressure plotted as a function of surface mass density (or surface concentration), exhibiting the virial regime giving rise to molecular weight dependence as in the osmotic virial regime in bulk solution, semi-dilute regime where the molecular weight dependence is obviated and state of monolayer collapse. M_n is the number average molecular weight and Γ^* is the overlap concentration of the highest molecular weight sample

density regime is very difficult. Hence, more efforts are directed at the semi-dilute regime where the surface pressure is more readily determined with accuracy but the molecular weight dependence for a polymeric system no longer holds. As with the bulk semi-dilute solution, any attempt to formulate an equation of state is abandoned. Instead, a surface density scaling is adopted for the static characterization [28, 29] for a variety of polymeric systems, following the scaling argument set forth by de Gennes [30]. Thus, the surface pressure dependence on the surface density is expressed as a power law, $\Pi \sim \Gamma^y$. Here, y is related to the 2D molecular weight exponent ν ($y = 2\nu/(2\nu - 1)$). The analog of the overlap concentration c* for bulk solution is the cross-over surface density Γ^*. The relevant length scale is therefore not the 2D root mean square end-to-end distance $R_2(0)$ but the distance between 2D density blobs ξ [31].

This approach leads to values of y that depend on the solvent quality on the interface. For the case where the interface serves as a good solvent, the mean-field theory predicts $\nu = 0.75$ [31] while the numerical calculation by Le Guillou and Zinn–Justin yields $\nu = 0.77$ [32], leading to predictions of y = 3 and 2.86, respectively. For cases where the interface gives rise to the theta sol-

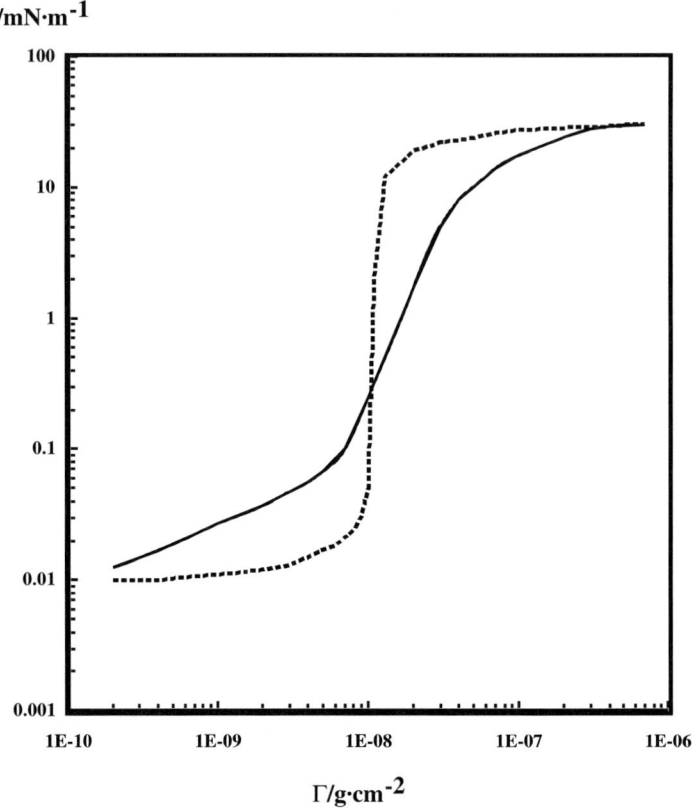

Fig. 2 Schematic contrast of isotherms in the good solvent limit (*solid curve*) and theta limit (*dashed curve*)

vent condition, the numerical method of Stephen and McCauley [33], yields $\nu = 0.505$ or $y = 101$. These two extremes are schematically depicted in Fig. 2.

Experimentally, good solvent conditions have been observed [22, 23, 27, 28, 34, 35]. On the other hand, none has been reported for the prediction of the theta condition, $y = 101$, whereas the prediction of poor solvent conditions giving rise to $y > 3$ has been reported. These all have $y \leq 20$ except for two; they are poly(methyl acrylate) at lower temperatures [34] and poly(dimethyl siloxane) [24]. Others have failed to reproduce them since. A caveat needs to be raised with these results. Since the semi-dilute regime is so narrow in Γ before the collapse state sets in whereby the power exponent is commonly deduced for a Γ range less than one full decade; hence, the Γ scaling is at best qualitative in the static characterization.

As seen above, the slope of the isotherm is useful for determining qualitatively the nature of the solvent quality of the interface. Additionally, the slope of the isotherm provides the lateral compressibility κ_λ or the lateral modulus

ε_s that are connected by a simple inverse relationship.

$$\kappa_\lambda = -\frac{1}{A}\left(\frac{\partial A}{\partial \Pi}\right)_T, \qquad (7)$$

$$\varepsilon_S = \kappa_\lambda^{-1} = -A\left(\frac{\partial \Pi}{\partial A}\right)_T. \qquad (8)$$

These are the 2D analogs of the bulk isothermal compressibility and bulk modulus, respectively. The scaling concepts introduced above clearly make sense in terms of the compressibility/elasticity as well. For polymers where the interface is a good solvent, the lateral modulus, sometimes called static dilational elasticity, is small whereas it becomes larger as the interface becomes poorer.

3
Capillary Wave Dynamics

An earlier episode in the history of surface wave effects is the observation by Benjamin Franklin [3] of calming ripples on Clapham Pond of nearly half an acre with a teaspoonful of oil. Although he did not explicitly estimate the thickness of the oil film, it turned out to correspond to roughly a monomolecular thickness [4]. This was no surprise, as seamen for centuries knew of this effect as well. The presence of the oil provides loss mechanisms in addition to the viscosity of water, which damp out the energy of these propagating gravitational waves. However, gravitational waves are not the only ones present. On the surface of any liquid, there is a continuous distribution of wavelengths associated with roughening due to spontaneous density fluctuations in the underlying liquid. These are called capillary waves or "ripplons", and their amplitude was determined only two decades ago to be rather small, 3–5 Å [36]. Thus, the interface of air/water is known to be molecularly smooth averaged over a macroscopic length scale (1 cm × 1 cm) and a time scale of hours. We now turn to examine in detail the dynamics of these waves. The theoretical underpinnings of the modern examination of the capillary waves are traced to a monograph by Levich [37] and the experimental efforts are traced to the pioneering laser light scattering studies by Katyl and Ingard [38, 39], simultaneously by Bouchiat et al. [40]. The capillary wave origin of surface laser light scattering was well established by Langevin and coworkers in the early 1970s [41–45], and she has subsequently lead the community in the study of surface-active substances on the interface, both experimentally and theoretically [46]. She coined the term, surface light scattering (SLS), and we will adhere to the designation.

One of the earliest theoretical studies in this area was work by Lord Kelvin [8, 47] who derived an expression for the frequency of a propagating wave on "an ideal pure liquid" (having zero viscosity), given as

$$\omega_0 = \sqrt{\frac{\sigma k^3}{\rho} + gk}, \quad (9)$$

where g is the gravitational constant, ρ liquid density and k wave vector, defined as $(2\pi/\lambda)$ with λ being capillary wavelength. Eq. 9 makes it clear the relative contributions of the wavelength dependent capillary and gravitational terms. The effects are illustrated in Fig. 3.

In passing we note that for $k > 10^4$ m^{-1}, the gravitational term contributes negligibly; the range of $k > 10^4$ m^{-1} is chosen in reference to the work to be presented later. As with hydrodynamics in general, the treatment with an ideal liquid is the starting point, and Eq. 9 predicts rather well for waves on pure liquids with a low viscosity including water. In addition, for a monolayer covered water surface the frequency deviates only by about 6% from that predicted by Eq. 9. A first-order correction to account for viscosity η on the capillary waves can be found in Langevin's monograph [46], called the corrected Kelvin equation, which is

$$\omega_0 = \sqrt{\frac{\sigma k^3}{\rho}} \left(1 - \frac{1}{2}\psi^{-3/4}\right), \quad (10)$$

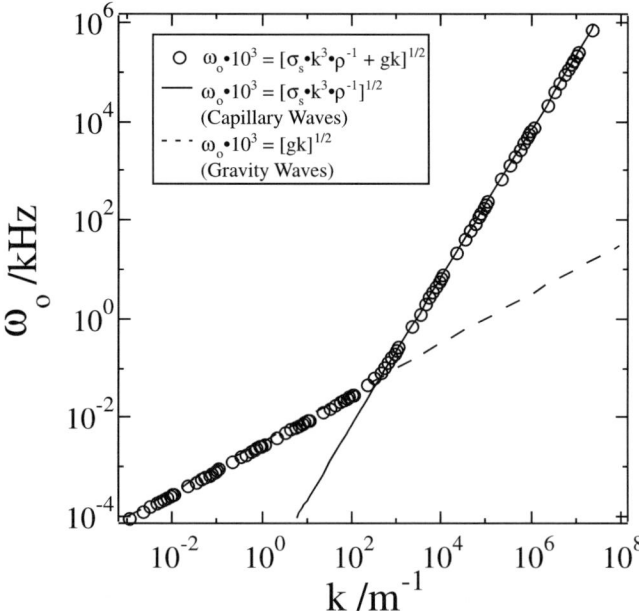

Fig. 3 Capillary and gravitational wave regimes, calculated for water at 25 °C ($\sigma = 71.97$ mN m^{-1}, $\rho = 997.0$ kg m^{-3} and $g = 9.80$ m s^{-2})

where ψ is a dimensionless subphase parameter which reflects the relative restoring and dissipating forces at the interface. First introduced by Bouchiat and Meunier [48] in solving for the power spectrum for surface light scattering from a pure liquid, ψ is defined as

$$\psi \equiv \frac{\sigma \rho}{4\eta^2 k}. \tag{11}$$

The modern resurgence in interest in capillary wave hydrodynamics, which started in the early 1950s, centers around the damping effects and the presence of a viscoelastic film between two fluids [37, 49–56]. All are more or less similar, in the assumptions invoked and the hydrodynamic theory used. The Lucassen-Reynders–Lucassen [55] and Kramer's [56] dispersion equations are essentially identical except Kramer ignores the gravity wave at the outset which is consistent with the wave vector range often used experimentally, and this is seen in Fig. 3.

We now turn to the exposition of the dispersion relation of Lucassen-Reynders–Lucassen for capillary waves at a fluid–fluid interface, given as

$$[\eta(k^* - m^*) - \eta'(k^* - m^{*\prime})]^2 = E \cdot S, \tag{12}$$

with

$$E \equiv \left[\eta(k^* + m^*) + \eta'(k^* + m^{*\prime}) + \frac{\varepsilon^* k^{*2}}{i\omega^*} \right], \tag{13}$$

and

$$S \equiv \left[\eta(k^* + m^*) + \eta'(k^* + m^{*\prime}) + \frac{\sigma^* k^{*2}}{i\omega^*} + \frac{g(\rho - \rho')}{i\omega^*} - \frac{\omega^*(\rho - \rho')}{ik^*} \right]. \tag{14}$$

In the above, the primed terms represent physical quantities in the upper phase, while the terms with superscript * represent complex quantities that are defined as

$$k^* \equiv k - i\beta, \tag{15}$$

$$\omega^* \equiv \omega_0 + i\alpha, \tag{16}$$

$$\sigma^* \equiv \sigma_d + i\omega^* \mu, \tag{17}$$

$$\varepsilon^* \equiv \varepsilon_d + i\omega^* \kappa, \tag{18}$$

and

$$m^* \equiv \left(k^{*2} + \frac{i\omega^* \rho}{\eta} \right)^{1/2}, \quad \mathrm{Re}(m^*) > 0. \tag{19}$$

For temporally damped waves, the complex wave vector k^* is just the real part k as the spatial damping coefficient β is equated to zero. Conversely, for spatially damped waves, the complex frequency ω^* is just the real frequency ω_0, as the temporal damping coefficient α is zero. Here, we restrict ourselves to the temporally damped waves, hence $k^* = k$. The other terms in the above represent the transverse modulus σ^*, consisting of an energy storage term of the dynamic surface tension σ_d and a loss term representing the imaginary part containing the transverse viscosity μ. Similarly, the complex lateral modulus ε^*, consists of the dynamic dilational elasticity ε_d and the corresponding loss term representing the imaginary dilational viscosity κ. The complex wave vector m^* gives rise to a measure of how deeply the surface waves penetrate into the underlying subphase liquid. Once all the definitions are provided, we return to the dispersion equation, Eq. 12. Despite its formidable appearance, it results in a good deal of simplification for temporally damped waves at air/liquid interfaces, since the upper phase can be neglected because of negligible air density ρ' and viscosity η' in comparison to those of the lower phase. Thus, the dispersion equation simplifies to

$$\eta^2(k-m^*)^2 = \left[\eta(k+m^*) + \frac{\epsilon^* k^2}{i\omega^*}\right]\left[\eta(k+m^*) + \frac{\sigma^* k^2}{i\omega^*} + \frac{g\rho}{i\omega^*} - \frac{\omega^* \rho}{ik}\right]. \tag{20}$$

In this article, we adopt Eq. 20 as the dispersion equation to analyze the dynamics of capillary waves on the air/water interface with polymeric monolayers.

We now turn to qualitative illustrations of capillary wave dynamics for the purposes of conveying the physics of the problem. On a pure liquid, surface tension is responsible for wave propagation at the surface in contact with air. As spontaneous density fluctuations induced by thermal fluctuations take place within a finite volume element of the liquid, they distort the surface causing an increase in the surface area. Then, the surface tension acts to minimize the surface area perturbation. In doing so, energy dissipation takes place by virtue of a finite viscosity of the liquid because of the momentum transfer among the liquid layers. The rest is transmitted to adjoining volume elements as kinetic energy, resulting in the surface wave propagation with decreasing amplitude with time. This simple picture is significantly affected by the presence of a surface film, as additional energy loss is associated with viscoelastic responses within the film itself. Goodrich [52] (1962) suggested five possible modes of monolayer motion. The approach of Lucassen-Reynders and Lucassen [55, 57–61] uses the transverse mode and the longitudinal mode to account for the additional damping of the liquid surface. It should be noted that there is an unknown in-plane shear component present in the longitudinal mode as well, however, this should be negligible for liquid-like films as liquids are readily deformable under shear [62]. Hence, the capillary waves are neither purely transverse nor purely longitudinal. For this reason, the observed capillary wave frequency ω^* represents a combination of the two whereby the damping observed at the surface results from a resonant mode coupling between the transverse and longitudinal modes.

We now turn to illustrate the mode coupling by plotting general solutions to the dispersion equation, first put forth by Hård and Neuman [63], and shown in Fig. 4. We first examine the limiting behaviors. The figure represents the temporal damping coefficient α plotted against the real capillary wave frequency ω_0. It stands for a specific solution at a specified wave vector on the water surface at 25 °C. The isopleths of constant dilational elasticity ε_d are drawn in radial solid curves emanating from point V, which will be explained shortly. The corresponding isopleths of constant dilational viscosity κ are drawn in circular dashed curves, emanating from a solid curve designated as VI which is the isopleth of $\varepsilon_d = 0$. We should dwell on these limits at this point because this gives the basis for the non-monotonic appearance of the experimental frequency and damping coefficient with respect to surface mass density Γ and surface pressure Π.

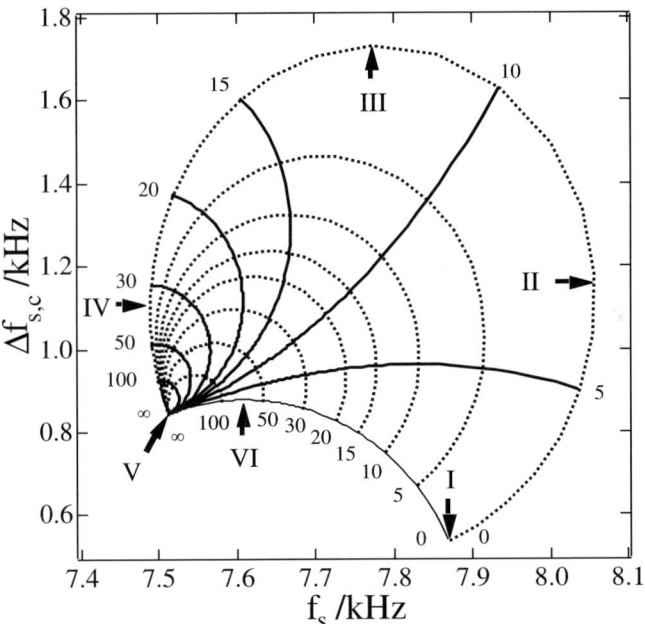

Fig. 4 General solution for the dispersion equation on water at 25 °C. The damping coefficient α vs. the real capillary wave frequency ω_0, for isopleths of constant dynamic dilation elasticity ε_d (*solid radial curves*), and dilational viscosity κ (*dashed circular curves*). The plot was generated for a reference subphase at $k = 32\,431\,\text{m}^{-1}$, $\sigma_d = 71.97\,\text{mN m}^{-1}$, $\mu = 0\,\text{mN s m}^{-1}$, $\rho = 997.0\,\text{kg m}^{-3}$, $\eta = 0.894\,\text{mPa s}$ and $g = 9.80\,\text{m s}^{-2}$. The limits correspond to I = Pure Liquid Limit, II = Maximum Velocity Limit for a Purely Elastic Surface Film, III = Maximum Damping Coefficient for the same, IV = Minimum Velocity Limit, V = Surface Film with an Infinite Lateral Modulus and VI = Maximum Damping Coefficient for a Perfectly Viscous Surface Film

Limit I

The state of a pure liquid without any viscoelastic film coverage is designated as I when $\varepsilon_d = 0$ and $\kappa = 0$. At this point, the frequency ω_o agrees with Eq. 10 while the damping coefficient α is given by the Stokes' equation [64] with a correction similar to the one used for the Kelvin equation:

$$\omega_o = \sqrt{\frac{\sigma k^3}{\rho}} \left(1 - \frac{1}{2}\psi^{-3/4}\right) \tag{10}$$

$$\alpha = \frac{2\eta k^2}{\rho}\left(1 - \frac{1}{2}y^{-1/4}\right) \qquad \alpha = \frac{2\eta k^2}{\rho}\left(1 - \frac{1}{2}\psi^{-1/4}\right), \tag{21}$$

Equation 21 is known as the corrected Stokes' equation. It takes into account the viscosity of the pure liquid to derive the damping coefficient, while the correction term arises due to the contribution of surface tension to the damping.

Limit II

Turning to the effects of viscoelastic films on a liquid surface, we note first that there are two obvious ones, i.e., purely elastic films and purely viscous films. For a purely elastic film ($\mu = 0$ and $\kappa = 0$), one starts at I, and proceeds around the outermost circular dashed curve ($\kappa = 0$) counterclockwise as the film elasticity increases. At point II, the capillary wave on a film-covered surface propagates faster than on a pure liquid. The physical basis is simple. The mode coupling at this point is such that the longitudinal mode is favored at the expense of the transverse mode. The wave frequency and damping under the conditions of Limit II are analogous but not identical to Eq. 10 and Eq. 21, respectively. From this limit to the last one, Limit V, one needs to introduce a set of corrections to the damping expressions given earlier by Stokes, Dorrenstein and Reynolds. The details of the corrections are provided by Esker in his doctoral thesis (1996). Skipping the details here, it suffices to merely state what they are and how they arise. For Limit II:

$$\omega_o = \sqrt{\frac{\sigma k^3}{\rho}} \left(1 + \frac{1}{387}\psi^{21/64}\right) \text{ (valid for } 210 < \psi < 490\text{)}, \tag{22}$$

and

$$\alpha = \frac{\sqrt{2}}{2}\left(\frac{\sigma\eta^2 k^7}{\rho^3}\right)^{1/4}\left(1 - \frac{2}{25}\psi^{5/27}\right). \tag{23}$$

In passing, we note that the applicability range of ω_o with respect to ψ is narrower than the other expressions that are valid over the full range of ψ values, customarily probed experimentally.

Limit III

The next significant limit is the case of the maximum damping coefficient. This arises when the frequencies of the transverse and longitudinal modes at the surface are equal. The frequency and damping coefficient are then given, respectively, as

$$\omega_0 = \sqrt{\frac{\sigma k^3}{\rho}} \left(1 - \frac{10}{17}\psi^{-9/16}\right) \tag{24}$$

and

$$\alpha = \frac{\sqrt{2}}{2} \left(\frac{\sigma \eta^2 k^7}{\rho^3}\right)^{1/4} \left(1 + \frac{4}{11}\psi^{-5/32}\right). \tag{25}$$

The limit of maximum damping coefficient at intermediate elasticities was first recognized by Dorrenstein [49] and corresponds to the maximum resonant coupling of the two modes at the surface, and the correction to the original Dorrenstein equation is prescribed by Esker. Lucassen-Reynders and Lucassen [55] provide some insight into the nature of this limit, which is illustrated in Fig. 5. The maximum in the damping coefficient signifies a change in the phase between the longitudinal and transverse modes from 90° at the pure liquid, Limit I, to 180° at the maximum damping coefficient, Limit III. Hence, the motion of the surface particles (monolayer) is at a 45° angle to the propagation direction of the capillary waves.

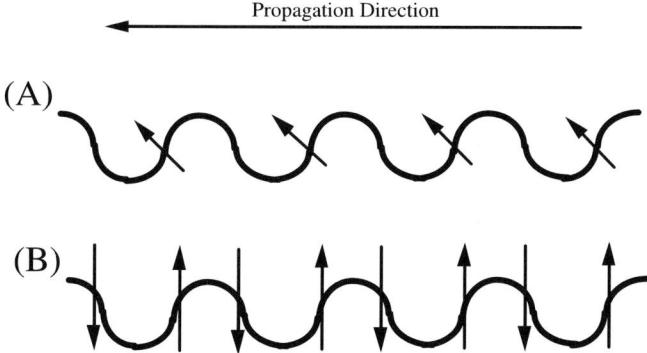

Fig. 5 Wave motion at maximum damping and infinite dilational elasticity. **A** Motion at the maximum damping coefficient where optimal resonant mode coupling implies that a surface fluid element moves at a 45° angle to the direction of wave propagation. **B** Wave Motion at infinite dilational elasticity, where the same element is only able to move in the transverse direction

Limit IV

Further increases in elasticity lead to the next limit of the minimum velocity for a perfectly elastic surface film. Here, the mode coupling leads to the case when the transverse motion is favored at the expense of lateral motion that is now practically stopped. Treating this case as a departure from the infinite lateral modulus case, Limit V, expressions for ω_o and α are given as follows.

$$\omega_o = \sqrt{\frac{\sigma k^3}{\rho}} \left(1 - \frac{10}{28}\psi^{-19/64}\right), \tag{26}$$

and

$$\alpha = \frac{\sqrt{2}}{4}\left(\frac{\sigma\eta^2 k^7}{\rho^3}\right)^{1/4}\left(1 + \frac{5}{3}y^{-25/121}\right). \tag{27}$$

Limit V

At the limit of infinite complex lateral modulus, still restricted to $\kappa = 0$, wherein we deal with the purely elastic limit such that $\varepsilon^* = \varepsilon_d \to \infty$, the longitudinal mode should cease. As such, the picture of the surface is one in which the surface elements are restricted to purely transverse mode, as depicted in Fig. 5. Thus, the frequency and damping coefficient are given as

$$\omega_o = \sqrt{\frac{\sigma k^3}{\rho}} \left(1 - \frac{1}{4}\psi^{-1/4}\right) \tag{28}$$

and

$$\alpha = \frac{\sqrt{2}}{4}\left(\frac{\sigma\eta^2 k^7}{\rho^3}\right)^{1/4}\left(1 + \frac{1}{2}\psi^{-1/4}\right). \tag{29}$$

Here, an earlier work by Reynolds [65] provided an expression for the damping of a surface film with an infinite lateral modulus, and Eq. 29 is called the corrected Reynolds equation, for it carries a correction as discussed above.

Limit VI

The limit is when $\varepsilon^* = i\omega^*\kappa \to \infty$ in the case of a purely viscous film with $\varepsilon_d = 0$. Putting it differently, the lateral modulus is a complex quantity, $\varepsilon^* = \varepsilon_d + i\omega\kappa$ (Eq. 18), hence the infinite limit must apply not only to the purely elastic surface film, $\kappa = 0$, but also to the purely viscous surface film, $\varepsilon_d = 0$. The dynamics are then closely related to those of a surface with the infinite lateral modulus. The expressions for both ω_o and α (with $\mu = 0$) are:

$$\omega_o = \sqrt{\frac{\sigma k^3}{\rho}} \left(1 - \frac{3}{25}\psi^{-2/11}\right), \tag{30}$$

and

$$\alpha = \frac{\sqrt{2}}{4} \left(\frac{\sigma \eta^2 k^7}{\rho^3}\right)^{1/4} \left(1 + \frac{25}{22}\psi^{-8/25}\right). \tag{31}$$

We conclude this discussion by noting that the role of κ is to suppress Limit III, the maximum damping coefficient limit. As is apparent in Fig. 4, the circular dashed curve decreases in radius of curvature with increasing surface viscosity, thus κ suppresses the maximum damping coefficient.

We come to some important points in the analysis of capillary wave dynamics through the polar plot profile displayed in Fig. 4. First are the assumptions that we invoke in the analysis. Although the details will be better clarified when we come to the experimental part dealing with the SLS method, we must at this point lay down the assumptions and how they are in part justified. Throughout the entire scheme of capillary wave analysis presented here, we make the following assumptions:

1. The transverse viscosity μ is negligibly small in the wave propagation;
2. $\sigma_d = \sigma_s$.

Addressing the first assumption, we display in Fig. 6 how μ affects the overall profile (by showing only $\mu = 0$ and $\kappa = 0$ traces) of the $\alpha - \omega_o$ polar plot. Basically, μ affects the damping coefficient α, whereas it influences ω_o negligibly. Our principal justification for the neglecting of μ is that it is not necessary to include it in the dispersion equation to reduce data at different wave vectors to a single polar plot. Since it affects the damping coefficient so sensitively, we could have readily detected its influence, if there were significant contributions. In short, we reduce Eq. 17, $\sigma^* = \sigma_d + i\omega^*\mu$ to $\sigma^* = \sigma_s$, which is a directly measurable experimental quantity in combination with the second assumption. This point will be further discussed in the experimental part. Our position is in part based on an experimental expediency of data reduction; quantitative specification of σ_d and μ does not aid the dynamic characterization much more than that under the assumption of $\sigma_d = \sigma_s$ and $\mu = 0$. On the other hand, the theoretical basis of neglecting μ is not clear-cut however small it may be [29, 66, 67]. Even experimentally there exists an alternative scheme to extract all four parameters, $\sigma_{d,\mu}$, ε_d and κ by analyzing the full time-domain signal of surface scattered light (see below). This was advocated by Earnshaw et al. [67] in 1990 and implemented by Henderson et al. [68] with poly(methyl methacrylate).

Next we come to the effect of surface tension on the dispersion relationship via the polar plot profile. This is shown in Fig. 7. Three features should be noted in the figure. First, the general profile of the polar plot does not

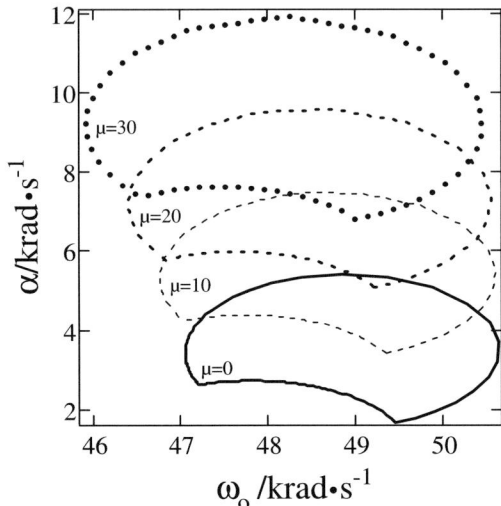

Fig. 6 The effect of transverse viscosity on the polar plot of Fig. 4. The damping coefficient, α, is plotted vs. the real capillary wave frequency, ω_o, for several different transverse viscosities (μ in the figure has units of 10^{-5} mN s m^{-1}). Only the isopleths for $\varepsilon_d = 0$ and $\kappa = 0$ are shown to give the outermost loop of Fig. 4. The plot was generated using the same condition as in Fig. 4, $k = 32\,431$ m^{-1}, $\sigma_d = 71.97$ mN m^{-1}, $\rho = 997.0$ kg m^{-3}, $\eta = 0.894$ mPa s and $g = 9.80$ m s^{-2}

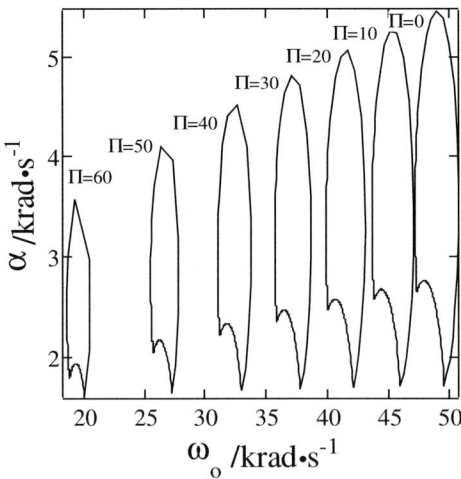

Fig. 7 The effect of surface tension on Fig. 4. The damping coefficient, α, is plotted vs. the real capillary wave frequency, ω_o, for several different surface tensions ($\sigma_d = 71.97 - \Pi$ mN m^{-1}). Only the isopleths for $\varepsilon_d = 0$ and $\kappa = 0$ are shown to give the outermost loop of Fig. 4. The plot was generated using $k = 32\,431$ m^{-1}, $\rho = 997.0$ kg m^{-3}, $\mu = 0$ mN s m^{-1}, $\eta = 0.894$ mPa s and $g = 9.80$ m s^{-2}

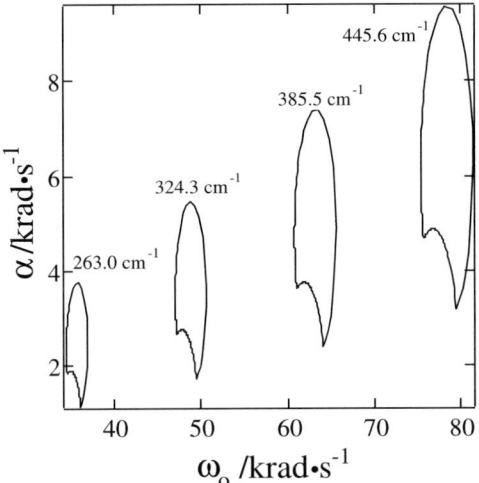

Fig. 8 The effect of wave vector on Fig. 4. The damping coefficient, α, is plotted vs. the real capillary wave frequency, ω_0, for several different wave vectors. Only the isopleths for $\varepsilon_d = 0$ and $\kappa = 0$ are shown to give the outermost loop of Fig. 4. The plot was generated using $\sigma_d = 71.97$ mN m^{-1}, $\mu = 0$ mN s m^{-1}, $\rho = 997.0$ kg m^{-3}, $\eta = 0.894$ mPa s and $g = 9.80$ m s^{-2}

vary with the surface tension unless it is depressed greatly, i.e., at high surface pressures. Second, it shows a monotonic decrease in the wave propagation rate as the surface tension increases as should be expected. Third, there also is a monotonic depression in the damping with increase in the surface tension, meaning that the capillary wave damping on a film-covered surface at high surface pressures can actually be less than the damping on a pure liquid surface, as was first noted in 1989 by McGivern and Earnshaw [69]. We hasten to add that we are dealing with the temporal damping.

Finally, we come to the effect of the wave vector k for a range used in the experiment to be presented. Figure 8 shows that the polar plot profile changes insignificantly over the range of k. Hence, it appears entirely justified in averaging and collapsing onto a single profile the results of α and ω_0 obtained at different wave vectors over the indicated range.

4
Surface Light Scattering Method

Light scattering as an experimental method to probe liquids and solutions has a long history since the demonstration by Tyndall in 1869 [70]. A vast literature exists on the subject of scattering in condensed matter (see for example, Fabelinskii [71]). Application to liquid surfaces began at the beginning of the 20th century with work by Smoluchowski [72], Mandelstam [73], Gans [74],

Adronov and Leontovich [75], Raman and Ramdas [76]. For bulk liquids, it took the advent of the laser to generate excitement in dynamics and critical phenomena through the scattered power spectrum or the time domain autocorrelation function [77–79]. As stated earlier, laser scattering from capillary waves on liquid surfaces was first performed in 1968–1989 by Katyl and Ingard [38, 39] and simultaneously by Bouchiat and Meunier [80, 81]. Langevin [82] provides a cogent and concise historical perspective on the subject in a monograph under her editorship. In this 1992 review, a significant body of work with pure liquid surfaces and surfactant-covered surfaces is presented except those concerning polymer-covered surfaces.

On the air/water interface, light scattering by capillary wave gives rise to a Doppler shifted and damping broadened power spectrum. The spontaneous capillary wave propagation with damping is just a surface analog of Brillouin scattering in bulk liquids [83] wherein spontaneous phonon propagation with damping gives arise to the Brillouin doublet. As referred to earlier, there are two schools of analysis, the frequency domain and the time domain. In the frequency domain, a heterodyne beat signal of the scattered field is acquired to produce the power spectrum, and its spectral profile together with its peak position are used to extract four or two viscoelastic parameters, σ_d, μ, ε_d and κ (Eqs. 17 and 18), or just ε_d and κ with the assumptions of $\sigma_d = \sigma_s$ and $\mu = 0$ together with the Lorentzian profile of the power spectrum. In the time domain, a heterodyne signal of the scattered field is obtained to give rise to the intensity autocorrelation function of exponential profile with amplitude oscillation corresponding to the wave propagation, which is then used to extract the four or two viscoelastic parameters. The first scheme with the frequency domain to extract two parameters is what we have adopted [84–87] and our results presented in this review are all from such a scheme.

Relative to the optical configuration of the instrumentation, one important element needs to be noted here. In the early configurations of Katyl and Ingard [38, 39], McQueen and Lundström [88], and Langevin [44], the provision

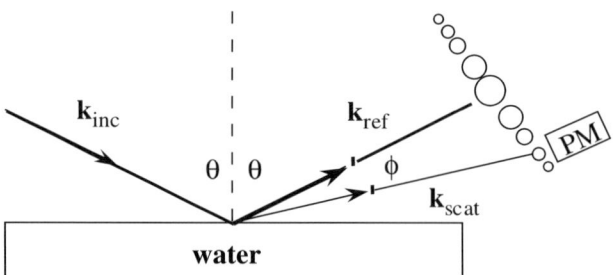

Fig. 9 Schematic diagram of the scattering geometry. The incident, reflected and scattered wave vectors are indicated by k_{inc}, k_{ref} and k_{scat}, respectively, while PM is the photomultiplier tube and θ and ϕ correspond to the incident and scattering angles

of a local oscillator to heterodyne with the scattered field was cumbersome and rather sensitive to mechanical vibration of the instrument [89]. A major innovation was introduced in 1976 by Hård et al. [90]. This was to insert a transmission grating in the incident beam such that the diffracted beams provided both a convenient and reproducible means of providing the local oscillator. Thus, the scattering wave vector is determined by the position of the diffraction beam. All optical configurations since 1976 follow this means to effect the heterodyne detection of the scattered light.

We present now how we deduce α and ω_0 from the scattering experiment. The incident light interacts with the propagating capillary waves and is scattered with a Doppler shift, analogous to Brillouin scattering in bulk liquid as stated earlier. Hence, the light interaction with ripplons is completely parallel with the case of the light interaction with phonons in bulk. The velocity of the capillary wave v is given by a rearrangement of Kelvin's equation, Eq. 9, as

$$v = \sqrt{\frac{g}{k} + \frac{\sigma_s k}{\rho}} = \sqrt{\frac{\sigma_s k}{\rho}} \quad \text{(Capillary Waves)} . \tag{32}$$

Hence, the frequency shift ν_{shift} from that of the incident light ν is expressed as

$$\nu_{shifted} = \nu_{incident}\left(1 \pm \frac{v}{c}\right) \quad \text{for} \quad v \ll c , \tag{33}$$

where c is the speed of light, and the observed frequency shift f_s in the power spectrum is given by

$$f_s = |\nu_{shifted} - \nu_{incident}| = \left(\frac{vk}{2\pi}\right) . \tag{34}$$

Since the scattered intensity power spectrum provides a folded version of the two directions, forward and backward shifts, sign $\pm$ disappears. The experimental optical configuration is illustrated in Fig. 11. It shows that the wave vector corresponding to nth order diffraction is defined as

$$k_{n,\|} = \frac{2\pi}{\lambda_\nu} \approx \frac{2\pi}{\lambda_L}\cos\theta \sin\phi_n \approx \frac{2\pi}{\lambda_L}\phi_n \cos\theta , \tag{35}$$

where λ_ν and λ_L are the wavelength of the capillary wave and laser, and θ and ϕ are the incident and scattering angle, respectively. An example is given for water at 25 °C for illustration purpose. Taking $\sigma_s = 71.97$ mN m^{-1} and $\rho = 997.0$ kg m^{-3} into Eq. 32 for $k = 26\,300$, $32\,430$ and $38\,550$ m^{-1}, leads to velocities of 1.378, 1.530 and 1.667 m s^{-1}. Using these numbers in Eq. 34 leads to frequency shifts of 5.768, 7.987 and 10.230 kHz, which are in turn confirmed experimentally, and this is displayed in Fig. 10. The instrument calibration is commonly effected with pure liquids such as water, ethanol and anisole [84], and validated with other liquids to confirm accurate determinations of surface tension and shear viscosity.

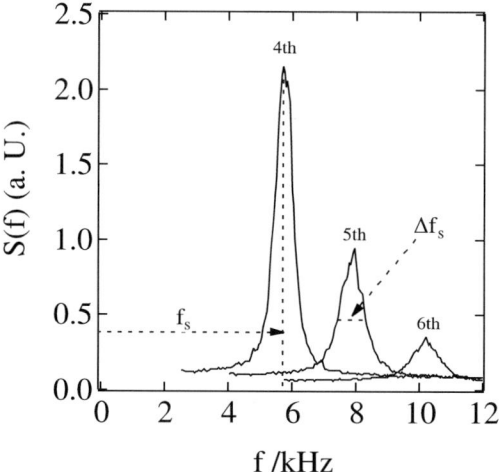

Fig. 10 Representative SLS power spectra from water at 25 °C. The three diffraction orders, 4th, 5th and 6th, correspond to $k = 26295$, 32431 and 38548 m^{-1}, respectively. The frequency shift for the 4th order is indicated by f_s while the full width at half-maximum intensity for the 5th order is indicated by Δf_s

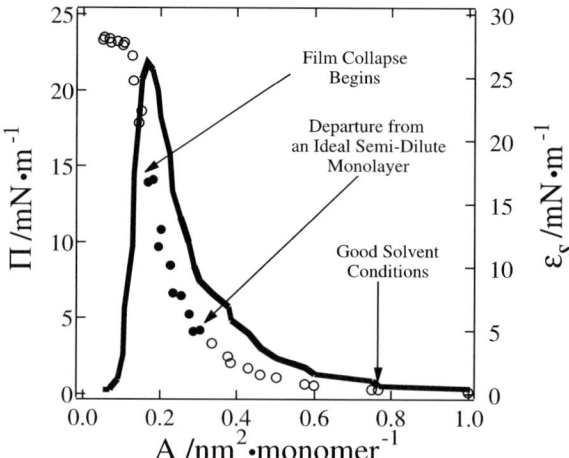

Fig. 11 Π and ε_s vs. A isotherms for PVAc on water at 23 °C. Surface pressure Π (represented by the *circles*) and the static elasticity ε_s (represented by the *solid curve*) are plotted as a function of area per monomer. The *open circles* at large A correspond to good solvent conditions. The *filled circles* correspond to the region of increasing ε_s as well as deviation from a power law scaling. The *open circles* at small A correspond to a collapsing film

Two other technical issues must be discussed briefly. In deducing the viscoelastic parameters through the dispersion equation from the power spectrum of SLS, it must be corrected for the instrumental broadening. The other is the approximation of observed spectral profiles by Lorentzian. SLS experimental results as we analyze them provide two quantities. They are the frequency shift f_s and the full width at half-maximum intensity of a spectrum Δf_s. The former is directly related to the capillary wave propagation rate, Eq. 34. The correction refers to the latter quantity, and it arises from the Gaussian profile of the incident laser beam, and is performed according to the procedure proposed by Hård and coworkers [90]. The instrumental width at half-maximum intensity is given as

$$\Delta f_i = \sqrt{2} \left(\frac{\Delta u_i \cos \theta_i}{R\lambda} \right) \left(\frac{d\omega_{o,k}}{dk} \right), \tag{36}$$

where θ_i is the incident angle, Δu_i is the full width at half-maximum intensity of the Gaussian beam profile of the ith diffraction order, λ is the wavelength of the laser, R is the distance between the detector and the interface and $\omega_{o,k}$ is the real frequency at wave vector k. Thus, the corrected full width at half-maximum intensity with Δf_i is given as

$$\Delta f_{s,c} = \Delta f_s - \frac{\Delta f_i^2}{\Delta f_s}, \tag{37}$$

where subscript s,c is to designate the instrumental broadening corrected spectral width. Turning to the Lorentzian approximation, we cite a detailed examination by Earnshaw and McGivern [91]. They note that the non-Lorentzian nature of the power spectrum accounts for less than a 0.25% error in f_s, and less than 1% in Δf_s, over the range of wave vectors probed here. These errors are much smaller than the magnitude of the instrumental correction and are also smaller than the random experimental errors which are typically on the order of 0.5–1% for f_s and 5–10% for $\Delta f_{s,c}$.

Now the two experimental quantities, f_s and $\Delta f_{s,c}$, are equated to the two in Eq. 16, $\omega_o = 2\pi f_s$ and $\alpha = \pi \Delta f_{s,c}$. With these two with the assumptions of $\sigma_d = \sigma_s$ and $\mu = 0$, the dispersion equation is solved for ε_d and κ. In the case of pure liquids, f_s and $\Delta f_{s,c}$ can be used to extract σ_s and η, the bulk viscosity. Additionally, the values of ε_d and κ can then be used to calculate the equivalent f_s and $\Delta f_{s,c}$ under the conditions of $\sigma_s = 71.97$ mN m^{-1}, $\mu = 0$ mN s m^{-1}, $k = 32431$ m^{-1}, $\rho = 997.0$ kg m^{-3} and $\eta = 0.894$ Pa s. These data can then all be plotted on a single polar plot of the kind shown in Fig. 4 after normalizing to a single wave vector with a specified set of subphase properties, i.e., surface tension, density and shear viscosity. Such a plot is then used to probe the trajectory of how ε_d and κ change with increasing Π. More importantly, one gains qualitative insight into the viscoelastic responses of monolayers placed at the interface at thermal equilibrium. It must be emphasized that we can deduce the parameters over a limited range, $\varepsilon_d < 60$ mN m^{-1} and $\kappa < 60$ mN s m^{-1}.

5
Polymer Systems

5.1
Homopolymers

5.1.1
Vinyl Polymers and Polyethers

Primary structural features of vinyl polymers are such that the pendant side chains can be hydrophilic whereas the backbone is just a polyethylene chain thus hydrophobic, rendering the entire chain structures amphiphilic. More picturesquely put, such a chain is said to have a greasy backbone decorated with a wet foot on every chain unit. This drives such vinyl chains to form monolayers on the air/water interface [92]. Prime examples are poly(vinyl acetate) and poly(methyl acrylate). Ries in 1961 [93] was the first to point out that vinyl polymer monolayers are horizontal in their chain conformation, whereas monolayers of small molecule surfactants, such as fatty acid, alcohol, phospholipids and glycerides, are vertical in conformation since each molecule stands on its polar head with a greasy tail pointed outward to the upper phase, air. Another kind of vinyl polymer chain, readily forming a monolayer on the air/water interface, has a more greasy backbone of polypropylene structure with the side chain equally hydrophilic or less so than the above examples. Prime examples of this kind are poly(methyl methacrylate) and poly(t-butyl methacrylate). A third class of polymers to consider is polyethers. Here, ether oxygen in the backbone is hydrophilic, whereas the connecting segment is hydrophobic. Examples are poly(ethylene oxide) and poly(butylene oxide) which is also called poly(tetrahydrofuran). Chemical structures of the three kinds are shown below.

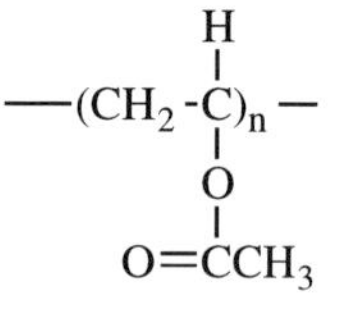

$$-(CH_2-\underset{\underset{\underset{O=CCH_3}{|}}{\overset{|}{O}}}{\overset{\overset{H}{|}}{C}})_n- \qquad -(CH_2-\underset{\underset{\underset{OCH_3}{|}}{\overset{|}{O=C}}}{\overset{\overset{H}{|}}{C}})_n-$$

Poly(vinyl acetate) (PVAc) 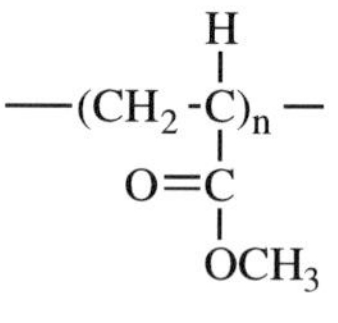   Poly(methyl acrylate) (PMA)

The conformational considerations lead us to expect that the monolayer thickness differs among the three classes that is accessible experimentally by the reflectivity techniques [94]. For instance, fatty acid monolayers have thicknesses on the order of 20 Å [95, 96], whereas the vinyl polymer monolayers of the first kind can have thicknesses of less than 10 Å [97]. On the other hand, the other class of vinyl polymers gives rise to thicker mono-

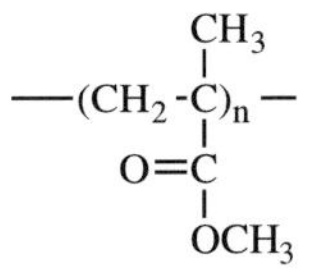

$$-(CH_2-\underset{\underset{OCH_3}{\overset{\|}{C}=O}}{\overset{CH_3}{|}})_n-$$

Poly(methyl methacrylate)

(PMMA)

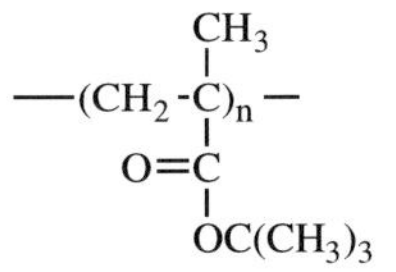

$$-(CH_2-\underset{\underset{OC(CH_3)_3}{\overset{\|}{C}=O}}{\overset{CH_3}{|}})_n-$$

Poly(t-butyl methacrylate)

(PtBMA)

$$-(OCH_2CH_2)_n-$$

Poly(ethylene oxide)

(PEO)

$$-(OCH_2CH_2CH_2CH_2)_n-$$

Poly(tetrahyrofuran)

(PTHF)

layers, for their chain conformations contain short loops into the subphase or upper phase, and also a significant fraction of water within the surface layer [68, 98–100]. Before proceeding to the discussion of the viscoelastic characterization of these homopolymers, the static properties of the monolayers are presented in the context of the scaling laws outlined in Sect. 2. The focus is on how the primary structures of the three classes affect the static properties. Recalling that the surface pressure is proportional to the surface density Γ^y, $\Pi \propto \Gamma^y$, where the scaling exponent y is given by

$$y = 2\nu/(2\nu - 1), \quad (38)$$

where ν is the molecular weight exponent. Making use of the inverse relationship of Γ and A, the surface pressure is expressed as

$$\Pi = C \cdot A^{-y}, \quad (39)$$

where C is a proportionality constant. With the definition of the static lateral modulus, Eq. 8, it can be recast by a simple proportionality with the constant y,

$$\varepsilon_s = y\Pi. \quad (40)$$

The determinations of ε_s from experimental $\Pi - A$ isotherms are illustrated with a polymer belonging to the first kind, poly(vinyl acetate), which happens to be the canonical example of polymer monolayer studies [10–12, 101, 102]. The isotherm at 23 °C is shown in Fig. 11, where A is expressed as area per repeating unit (monomer for short) of PVAc. On the same plot, ε_s is also given as a function of A. The Π-A isotherm of this sort for PVAc has a long track record and been reproduced repeatedly [35, 41, 84, 103–107]. The isotherm has been broken up into three distinct regimes for illustrative purposes. The first is the semi-dilute regime above the virial regime where the good solvent scaling behavior is observed; we will return to this point shortly. The next regime corresponds to the region over which the lateral modulus ε_s reaches a maximum and departure from the semi-dilute monolayer sets in. The last corresponds to the film collapse regime. Parenthetically, Ries and Walker [93] proposed with electron micrograph evidence that the collapse takes place via multilayer formation. However, in light of recent neutron reflectivity experimental analysis on the amount of water incorporated in a similar polymer monolayer, poly(methyl methacrylate) by Henderson et al. [68, 98] at all surface concentrations, a looped structure cannot be ruled out.

Continuing to use PVAc as the canonical example of a stable and easily reproducible polymer monolayer, we show how the two quantities, the static elasticity ε_s from Π-A the isotherm, and the corresponding ε_d deduced from the SLS experiment, compare and contrast with each other; for the time being, we defer to later the SLS results. This is shown in Fig. 12. Agreement between the two is remarkable up to respective maximum points. The observed deviation at higher Π is not expected since the monolayer state is no longer maintained, hence the static elastic responses in macroscopic scales are not likely to be the same as the dynamic response to spontaneous capillary waves.

Having shown how the static lateral modulus ε_s, also called static elasticity, is determined from Π-A isotherm via Eq. 8, we return to the three classes of homopolymers. Plots of ε_s against Π are shown in Fig. 13, where the re-

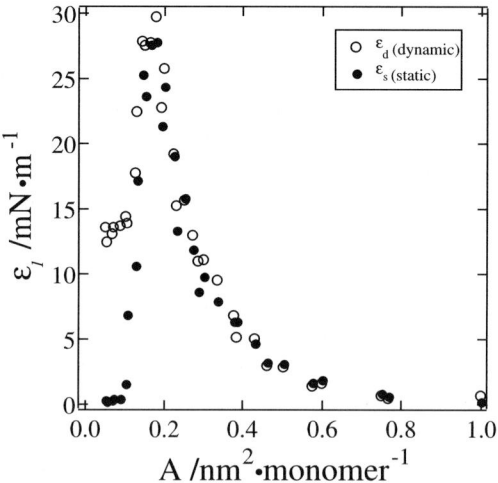

Fig. 12 ε_i vs. A for PVAc on water at 23 °C. Static and dynamic elasticities (averages of three wave vectors) are compared as a function of area per monomer. The agreement between the two is quantitative up to the maximum elasticity (≈ 28 mN m^{-1}). After this point as the film collapses, the dynamic values are higher than the static values and approach a limit of 13 ± 2 mN m^{-1}. Estimates of the errors are about the same as in Fig. 17

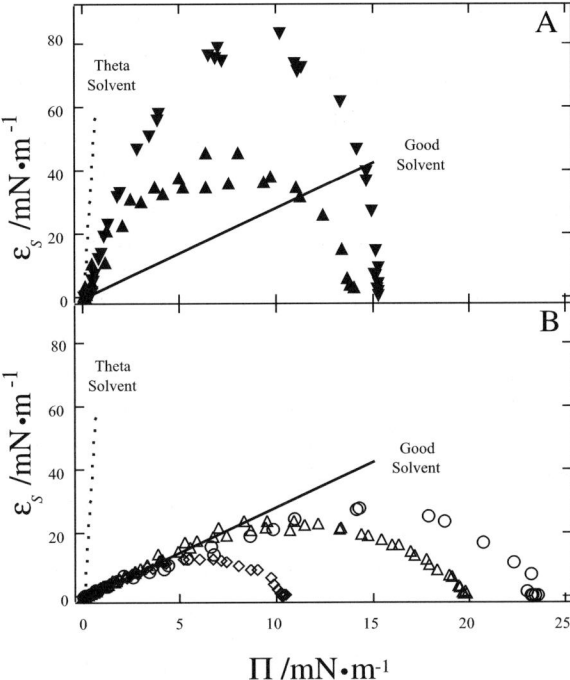

Fig. 13 $\varepsilon_s - \Pi$ for polymeric monolayers at A/W. The symbols correspond to data for PtBMA (▼), PMMA (▲), PVAc (○), PMA (△), PEO (◇), PTHF (□). The initial slopes in both are drawn in for the good solvent limit (*solid line*) and the theta limit (*dashed line*)

sults for six polymers are displayed. The initial linear region stands for the semi-dilute regime. As referred to above, the maximum ε_s in each case is observed well beyond the point when the isotherms depart from the semi-dilute regime. The initial slopes in the plots provide the scaling exponent y, and clearly a division into two groups emerges. The good solvent condition represented by solid lines applies to PVAc, PMA, PEO and PTHF with y ≈ 3 (2.86 to be exact), whereas a poor solvent condition (not quite theta condition indicated by dashed lines) with y ≈ 18, applies to PMMA and PtBMA. Clearly, methacrylate polymers with a polypropylene backbone form different monolayers than those with the polyethylene backbone or polyether. We truncate further presentation of the static results and proceed directly to the dynamics by SLS.

We present the viscoelastic characterization of these polymers as deduced from SLS. By means of the scheme outlined with Fig. 4, we demonstrate in Fig. 14 that two groups of polymers are well differentiated in terms of their "polar profile", that is their progression of ε_d and κ with increasing Π. In the figure, those under the good solvent condition with y ≈ 3, PVAc, PMA, PEO and PTHF, are shown in (A) and those under the poor solvent condi-

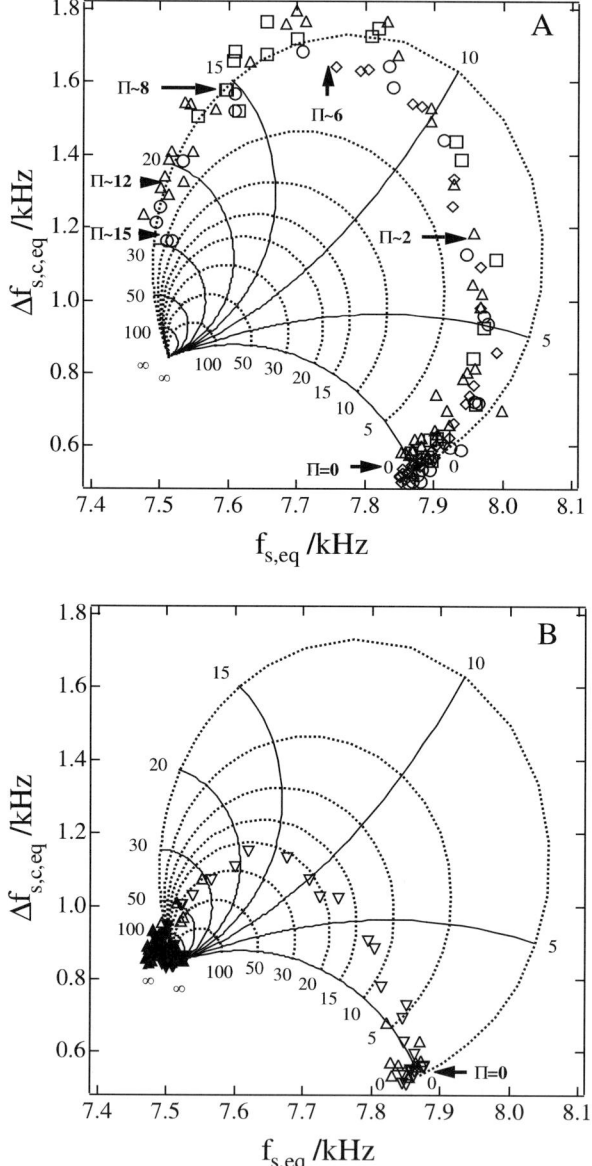

Fig. 14 $\Delta f_{s,c,eq}$ vs. $f_{s,eq}$ for PEO, PTHF, PMA and PVAc up to $\varepsilon_{s,max}$ for all four polymers, with the symbols identical to those in Fig. 13 (**A**). The same plots for PMMA and PtBMA are shown in (**B**), where the *open symbols* stand for $\Pi < 2$ mN m^{-1} and the *filled symbols* for $\Pi > 2$ mN m^{-1}. The *solid and dashed curves* are the same as in Fig. 4, and the surface pressure increases counterclockwise, starting from $\Pi = 0$, Limit I, in Fig. 4. PMMA shows a discontinuous change with can be explained by the coalescence of PMMA patches existing as a heterogeneous film prior to the monolayer state. Error bars, not shown for clarity, are 0.5% and 5% for $f_{s,eq}$ and $\Delta f_{s,c,eq}$, respectively

tion with y ≈ 18 are shown in (B). Thus, qualitatively it can be surmised that the first group shows behavior approximating purely elastic film with almost a constant range of viscosity, $0 \leq \kappa \leq 5 \times 10^{-5}$ mN s m^{-1}, whereas the second

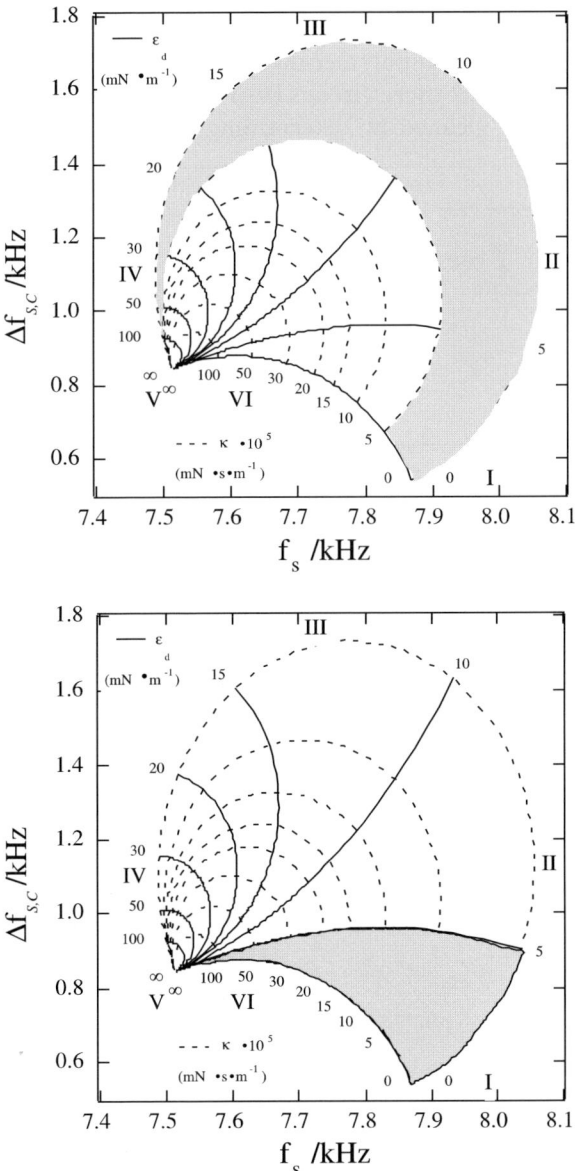

Fig. 15 $\Delta f_{s,c}$ vs. f_s for a roughly purely elastic film (*upper*) and a nearly perfectly viscous film (*lower*) are represented by the *shaded area*, where the conditions are the same as those of Fig. 4

group shows behavior rather akin to purely viscous film with a constant range of dynamic elasticity, $0 \leq \varepsilon_d \leq 5$ mN m^{-1}. Such a differentiation constitutes a major finding in this analysis scheme, particularly in terms of concordance between the static and dynamic properties. In the context of the classification given by Crisp in the early 1940s, the expanded polymer monolayers spread on water as the good solvent environments are more elastic than viscous. On the contrary, the condensed polymer monolayers spread on water as the poor solvent environments are more viscous than elastic. This sort of summary analysis may be better gleaned by determining where the data spread on the

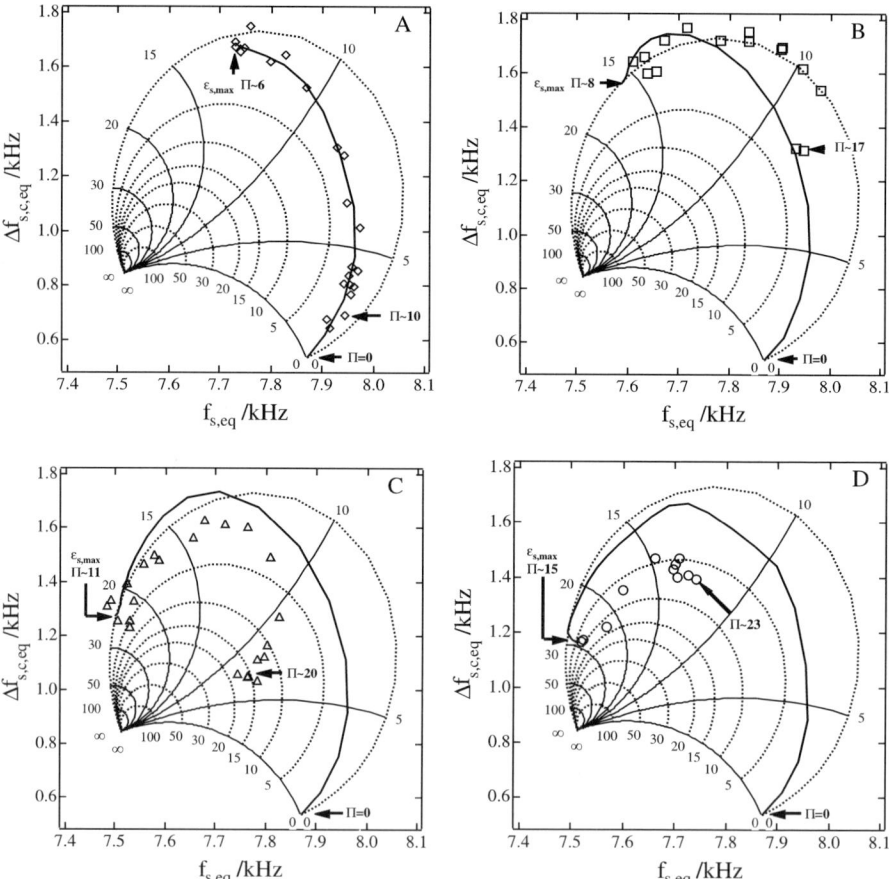

Fig. 16 $\Delta f_{s,c,eq}$ vs. $f_{s,eq}$ for PEO (**A**), PTHF (**B**), PMA (**C**) and PVAc (**D**) are plotted from $\epsilon_{s,max}$ the chain collapse, and the data shown in Fig. 14A are represented by *solid curves* for clarity, that are drawn from Limit I to $\varepsilon_{s,max}$. In each case P increases clockwise from indicated $\varepsilon_{s,max}$ to $\Pi \approx 10$ for PEO (**A**), $\Pi \approx 17$ for PTHF (**B**), $\Pi \approx 20$ for PMA (**C**), and $\Pi \approx 23$ for PVAc (**D**), and the unit of Π is given in mN m^{-1}. Error bars are the same as given in the legend of Fig. 14, but not shown for clarity

polar plot. Such a broad stroke analysis is exemplified with Fig. 15, where the shaded area roughly stands for either purely elastic monolayer or purely viscous monolayers, both within certain ranges of ε_d and κ.

Although we have skipped a great deal of technical details pertaining to the results in Fig. 14, we must note two points here. First, the data are av-

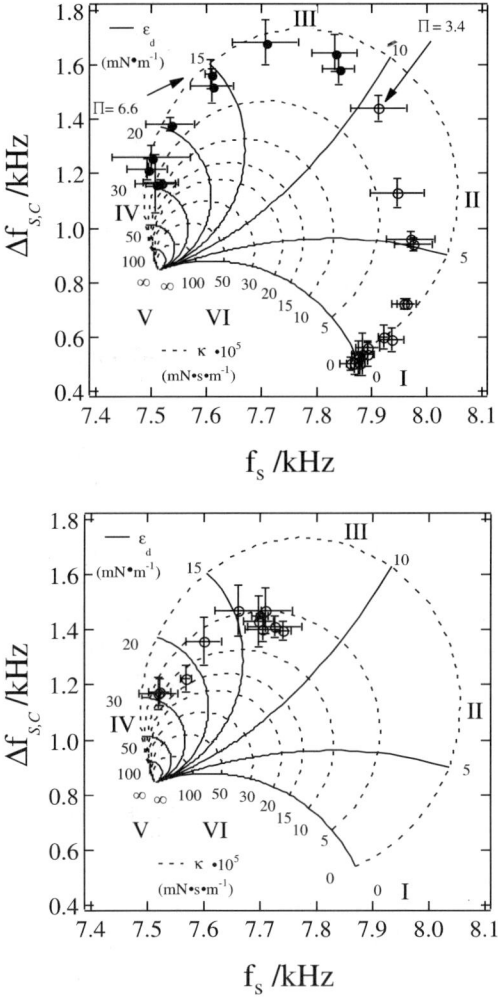

Fig. 17 $\Delta f_{s,c}$ vs. f_s for PVAc on water at 23 °C for $\Pi < 15$ mN m^{-1} (**A**) and $\Pi < 15$ mN m^{-1}, which is the same as Fig. 16D but with error bars. Values for different surface pressures have been normalized to the conditions and limits of Fig. 16 with 1 standard deviation confidence intervals for the average of three wave vectors. Surface pressure increases counterclockwise in **A** and clockwise in **B**. Complete collapse of the film occurs at a surface pressure around 23.5 mN m^{-1}. On the plot B, this corresponds to the cluster of points at ε_d around 13 ± 2 mN m^{-1} and $\kappa \cdot 10^5 \approx 6 \pm 2$ mN s m^{-1}

eraged over different wave vectors k over a relatively small accessible range, from 264 cm^{-1} (4th order diffraction) to 387 cm^{-1} (6th order diffraction). Second, with the good solvent monolayers, the data progression with increasing Π is such that they start at Limit I, move counterclockwise toward Limits III and IV and then loop back clockwise from $\varepsilon_{d,max}$ either along the same trajectory or move toward the interior of the plot with higher viscosities. These are illustrated in Fig. 16. In Fig. 17, we divide both paths, counterclockwise and clockwise progression with Π for PVAc (Fig. 16D), to show the details of paths with error bars. Our rationale for not dwelling much on the clockwise trajectory, however, arises from the fact that this regime entails onset of monolayer collapse, hence is no longer susceptible to the analysis scheme based on the dispersion equation. The situation is far more complex with the monolayers under the poor solvent conditions, all approaching Limit V.

5.1.2
Binary Monolayer: Side Chain Length Effect

Chen et al. [108] reported that poly(vinyl stearate) (PVS) could be studied only in the dilute regime ($\Pi < 5$ mN m^{-1}), as films compressed to higher surface pressures showed continuous relaxation in the surface pressure. Hence,

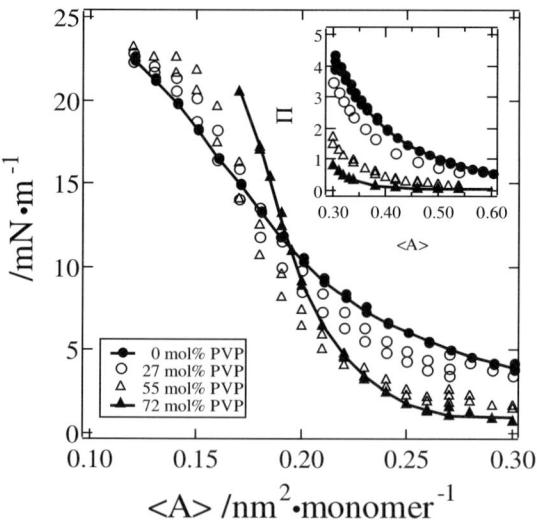

Fig. 18 Π – $<A>$ Isotherms for PVP/PVAc binary monolayers on water at 25 °C. Surface pressure Π for a variety of poly(vinyl palmitate)/poly(vinyl acetate) mixtures as defined in the plot are shown as a function of area per monomer. Surface concentration was controlled by step-wise compression. The incorporation of PVP, which does not form stable monolayers alone, condenses the film and also increases the instability of the film. $<A>$ = average area per monomer

a shorter side chain was examined with poly(vinyl palmitate), but it too gave rise to unstable films. Thus, the side chain length effect on vinyl polymer is examined by a binary system of PVP and PVAc [109]. Examples of isotherms from stepwise compression are shown in Fig. 18. Increasing the PVP composition causes the film to become more condensed, while the film stability is decreased. The film instability is well shown in Fig. 19 where both compression and stepwise addition were employed for a 72 mol % PVP monolayer. The isotherms are essentially identical up to $\Pi \approx 5$ mN m^{-1}, however, beyond this point $\Pi_{\text{compression}}$ was always greater than Π_{addition}, a clear signature for long relaxation times of the film upon compression. Thus, $\Pi \geq 5$ mN m^{-1}, we are no longer in thermodynamic equilibrium if we proceed with the compression method.

Turning to the viscoelastic properties of these binary monolayers, a simple trend is confirmed. The larger is the content of PVP, the more viscous becomes the monolayer, starting from the nearly elastic monolayer of PVAc. We show the trend by means of the polar plots at two compositions, 27 mol % and 72 mol % but not an intermediate third composition; the latter case is shown only for results obtained with the successive addition method. They are shown in Figs. 20 and 21. Although the plots appear to resemble that of PVAc, Figs. 14A and 16D, several key differences emerge. For the

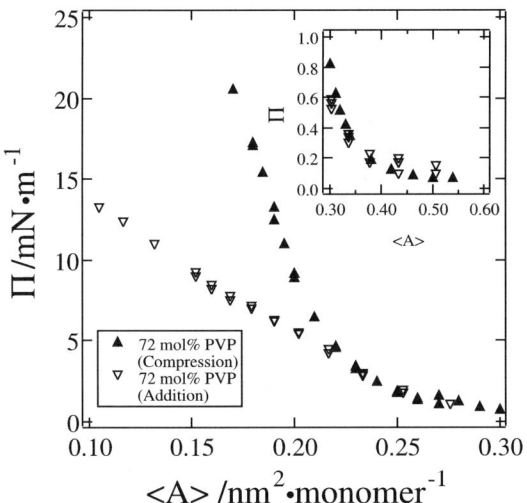

Fig. 19 $\Pi - <A>$ isotherms for 72 mol % PVP/PVAc mixed monolayers on water at 25 °C by stepwise addition and compression. Surface pressure Π for 72 mol % poly(vinyl palmitate)/poly(vinyl acetate) binary mixture as a function of area per monomer. The surface concentration was controlled as noted in the plot. For the stepwise-addition technique, lens formation was observed in the region where the two techniques differ for $\Pi > \approx 10$ mN m^{-1}. For the plot, the mixture required stabilization times considerably longer than the 1–2 hours allowed between points to form "equilibrium" films. $<A> =$ average area per monomer

72 mol % PVP mixtures, there is a significant increase in κ around Limit II, when $<A> \approx 0.38$ nm^2 monomer^{-1}, $\Pi \approx 0.2$ mN m^{-1}, $\varepsilon_d \approx 6$ mN m^{-1} and $\kappa \cdot 10^5 \approx 6$ mN s m^{-1}. Additionally, the maximum damping coefficient seems to be depressed which is consistent with an increase in κ. Another difference is the decrease in Π at the maximum damping coefficient at Limit III: $\Pi_{PVAc} > \Pi_{27\%PVP} > \Pi_{55\%PVP} > \Pi_{72\%PVP}$, corresponding to $\Pi = 4.8, 4.3, 1.5$ and 0.35 mN m^{-1}. This is similar to the trend observed for PVAc, PEO and PtBMA, again suggesting poorer hydrodynamic coupling between the monolayer and liquid motion. Finally, there is a dramatic change in the collapse behavior of these films. As mol% of PVP increases, the collapse behavior changes from that of pure PVAc, where the maximum elasticity roughly corresponds to Limit IV, the minimum velocity of a perfectly elastic surface, and collapse proceeds back along a path slightly more viscous than a perfectly elastic film to lower elasticity values (Fig. 18D). Rather than collapsing back to lower elasticities, the higher PVP content films proceed onto the infinite lateral modulus, Limit V, forming an essentially infinitely rigid film. This is attributable to the enhanced lateral interactions of the packed hydrophobic side chains which lead to higher film viscosities. For the case of 72 mol % PVP,

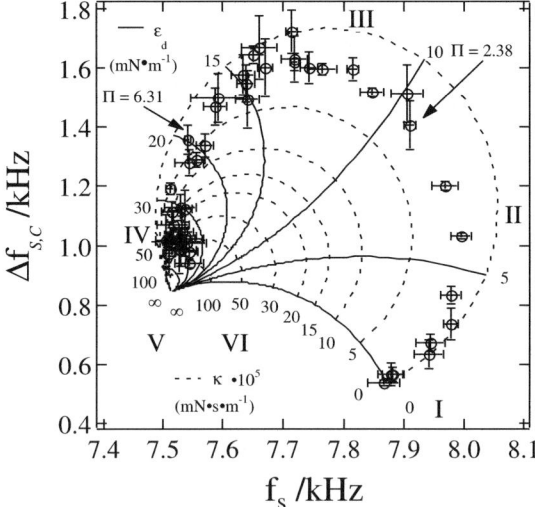

Fig. 20 $\Delta f_{s,c}$ vs. f_s for PVP/PVAc mixed monolayers on water at 25 °C (27 mol % PVP). Values for different surface pressures and wave vectors have been normalized and averaged to the conditions and limits of Fig. 4 with one standard deviation error bars. Surface pressure increases counterclockwise around the contours. Departure from pure liquid dynamics occurs at $<A>$ is less than around 0.6 nm^2 monomer^{-1}, and $\Pi > 0.6$ mN m^{-1}. The mixture approaches Limit II(A, Π, ε_d, $\kappa \cdot 10^5$) ≈ 0.36 nm^2 monomer^{-1}, 2 mN m^{-1}, 7 mN m^{-1}, 2 mN s m^{-1}, Limit III(A, Π, ε_d, $\kappa 10^5$) ≈ 0.27 nm^2 monomer^{-1}, 4.3 mN m^{-1}, 13 mN m^{-1}, 2 mN s m^{-1}, and Limit IV(A <, Π >) ≈ 0.17 nm^2 monomer^{-1}, 13 mN m^{-1} of a perfectly elastic surface film

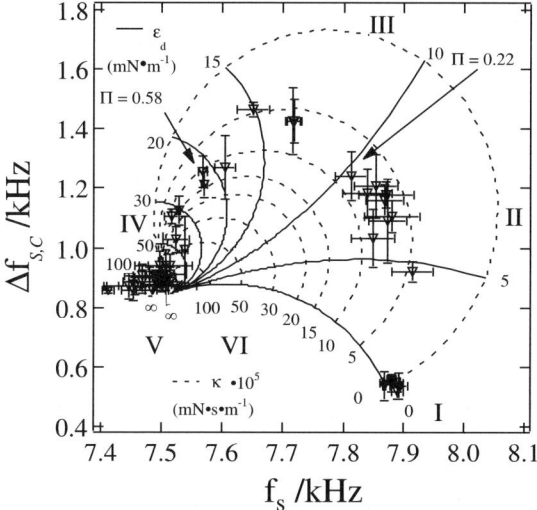

Fig. 21 $\Delta f_{s,c}$ vs. f_s for PVP/PVAc mixed monolayers on water at 25 °C (72 mol % PVP by successive addition). Values for different surface pressures and wave vectors have been averaged and normalized to the conditions and limits of Fig. 4 with 1 standard deviation error bars. Surface pressure increases counterclockwise around the contours

there is actually significant deviation even from this limit as the data actually fall outside the contours. Here, μ may be significant, or $\sigma_s \neq \sigma_d$, however, it is clear this region corresponds to a collapsing film as lens formation occurs during spreading, and the long stabilization times in this region would make further study difficult.

A binary monolayer of another kind was examined by means of the mechanical relaxation technique in conjunction with SLS by Rivillon et al. [110]. The system was PVAc and poly(p-hydroxystyrene) (P4HS). They have studied over a broad range of frequency (0.1 mHz–200 kHz) and finer composition resolution. Their conclusion is somewhat the same as our example of the PVP-PVAc system, but they can advance further details of the composition-dependent relaxation modes, after having established that the P4HS monolayer is in a poor solvent condition with the same set of two techniques [111].

5.1.3
Temperature Dependence

Although an accessible temperature range is restricted by complication of subphase water evaporation, there have been some reports on temperature-dependent investigations with homopolymers. Yoo and Yu [105] studied PVAc and poly(n-butyl methacrylate)(PnBMA) as the prototype of good solvent and poor solvent cases. Both were over a range of 15 °C, 10–25 °C and 15–30 °C respectively, and PnBMA required the inclusion of μ assuming that ε_d and κ did not depend on wave vector k, whereas such was not the case with PVAc. The transverse viscosity of PnBMA showed an anomalous temperature dependence, and was interpreted in terms of the relaxation model of Mann and Du [112]. On the other hand, Rubio's group have examined the temperature dependence over 1–25 °C with four different techniques [113, 114] including SLS. They found a transition temperature from a soft solid-like state at low temperature to a fluid state that is commonly observed.

5.2
Copolymers

5.2.1
Alternating Copolymers

We now turn to an intriguing alternating copolymer system [115]. It is the system of poly(1-alkylene-co-maleic acid), thus having the same hydrophilic head group while varying the length of hydrophobic side chain, from n-butyl to n-hexadecyl. The polymer system is designated as PXcMA, where X varies from 1-hexylene to 1-octadecylene. Thus, the side chain R varies from the n-butyl to n-hexadecyl group with $4 \leq m \leq 16$. It is synthesized by copolymerization of α-olefin with maleic anhydride, and subsequently hydrolyzed to a maleic acid moiety in the chain, constituting the hydrophilic component. We present the results for four polymers: they are (1) $m = 4$, the starting α-olefin is 1-hexene, hence designated as PHcMA, (2) $m = 6$, α-olefin=1-octene, POcMA, (3) $m = 8$, α-olefin = 1-decene, PDcMA, and (4) $m = 16$, α-olefin = 1-octadecene, PODcMA. The central focus here is to probe how the side chain length R influences the viscoelastic characteristics of the monolayers of these polymers.

As in the earlier section, we briefly present the static properties first. The results obtained with pH 2 water (with 10 mM HCl) as the subphase are displayed in Fig. 22, where Π-A and ε_s – A isotherms for all four polymer monolayers are plotted together in each. With the exception of PHcMA, the Π-A isotherms of the polymer monolayers are nearly identical and they are consistent with the expected isotherms of condensed films. Each isotherm is shown in two regions: the dilute regime where $\Pi \leq 1$ mN m^{-1} (open sym-

$$-(CH_2-\underset{\underset{R}{|}}{CH}-\underset{\underset{\underset{O}{\|}}{COH}}{CH}-\underset{\underset{\underset{O}{\|}}{COH}}{CH})_n$$

$$R: -C_mH_{2m+1}$$

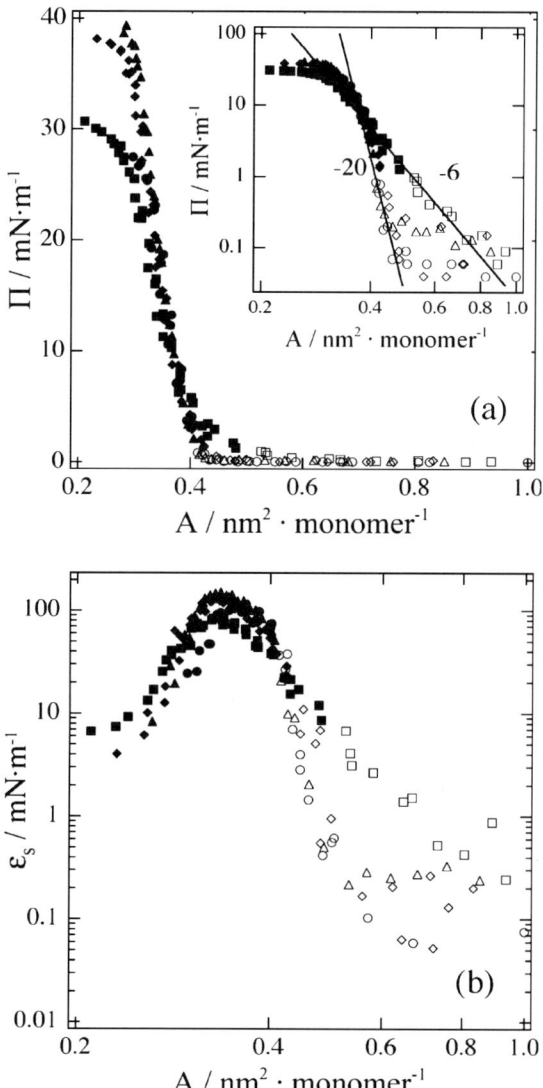

Fig. 22 (a) $\Pi - A$ Isotherms and (b) $\varepsilon_s - A$ for PXcMA polymers on 0.01 M HCl at 25 °C, both in a double logarithmic plot. *Squares* are for PHcMA, *diamonds* for POcMA, *triangles* for PDcMA, and *circles* for PODcMA. *Open symbols* are used for the dilute regime ($\Pi \leq 1$ mN m^{-1}) and *filled symbols* for the semi-dilute regime ($\Pi > 1$ mN m^{-1})

bols) and the semi-dilute regime where $\Pi > 1$ mN m^{-1} (filled symbols). The critical scaling exponent y in Eq. 39 for PHcMA is 6, while y for the others is about 20. This suggests that all three monolayers of POcMA, PDcMA and PODcMA, are placed in a poor solvent environment, whereas PHcMA is in a slightly better solvent, but is not at the good solvent limit. The static elas-

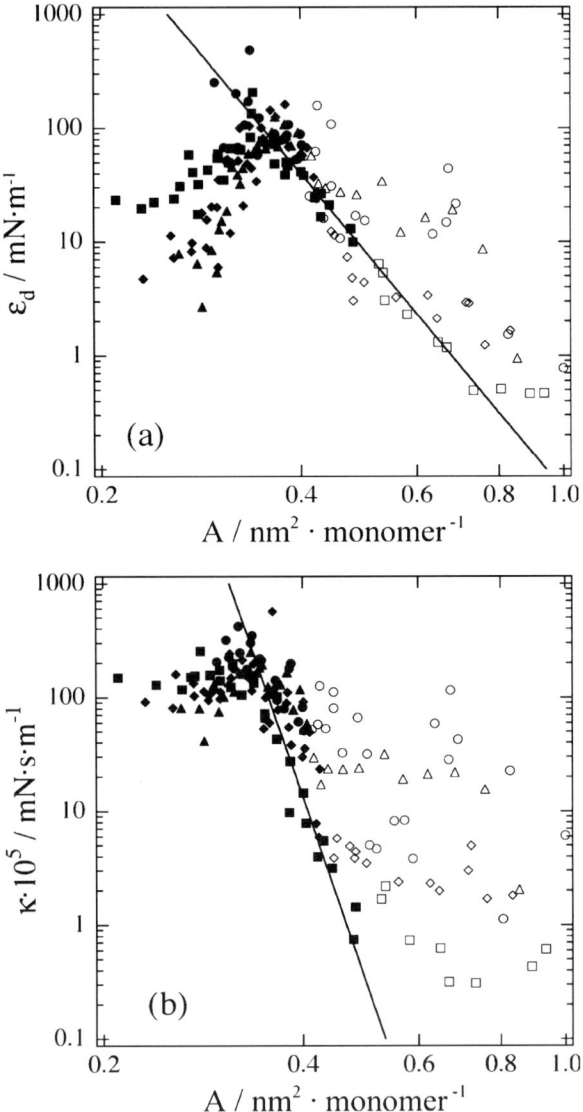

Fig. 23 (a) $\varepsilon_d - A$ and (b) $\kappa - A$ for PXcMA polymers on 0.01 M HCl at 25 °C. The *symbols* are the same as Fig. 22. The *solid line* is drawn over the data for PHcMA. The values represent an average over three wave vectors and error bars are omitted for clarity, but they amount to about 10 and 15%, respectively, for ε_d and κ

ticity ε_s values for PXcMA polymers are distinguishable in the dilute regime but collapsed together in the semi-dilute regime, again with the exception of PHcMA. The critical area A^* when the elasticity reaches a maximum, is 0.33–0.35 nm² for all PXcMA polymers and the calculated $\varepsilon_{s,max}$ value ranges from 80–140 mN m^{-1}. All values of A^* for PXcMA are less than twice the A^* of 0.18 nm^2 for PVAc, taken as the canonical example of surface-active vinyl polymer, whereas $\varepsilon_{s,max}$ is higher than those of the vinyl polymers [29]. The basis of this comparison arises from the fact that the PXcMA backbone repeating unit (monomer) has four carbons while that of the vinyl polymers has two carbons. This suggests that PXcMA polymers laterally pack more efficiently than the vinyl polymers.

Presenting the viscoelastic characteristics of the monolayers, we show in Fig. 23 ε_d vs. A and κ vs. A, all deduced by solving for the dispersion equation under the same set of assumptions. Although the differentiation between the dilute and semi-dilute regimes by open and filled symbols appear somewhat arbitrary, we will clarify this point shortly. In general, the behavior of both parameters, ε_d and κ, parallel that of ε_s. It is clear that they increase with greater lateral packing (decreasing A) up to $A^* \approx 0.32$ nm^2/monomer and then decrease from that point on. In the dilute regime (open symbols), longer side chain polymers produce only slightly higher ε_d values than those with shorter side chains. On the other hand, κ shows a clear side-chain dependence; the longer the side chain, the higher the value of κ. In the semi-dilute regime (filled symbols), however, all four appear to be the same.

In Fig. 24 a polar plot with all four polymers is displayed. PHcMA (C4 side chain) follows the profile of an almost purely elastic surface film whereas chains with longer side chains follow that of progressively more viscous pro-

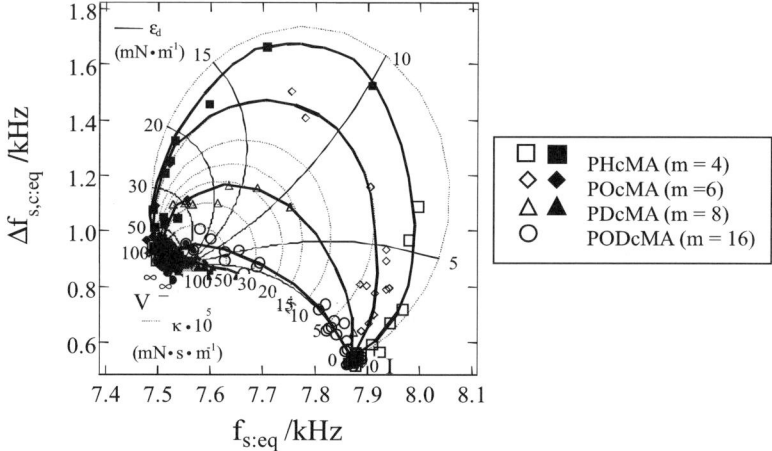

Fig. 24 Polar plot for PXcMA polymers on 0.01 M HCl at 25 °C. The *symbols* are the same as Fig. 22, and calculated based on the reference state that $k_{ref} = 324.3$ cm^{-1}, $\eta_{ref} = 0.894$ cP, $\rho_{ref} = 0.997$ g/cm^3, $\sigma_{d,ref} = 71.79$ mN/m, and $\mu_{ref} = 0$ mN s/m (water at 25 °C). The *solid curves* correspond to a constant value of ε_d while the *dashed loops* correspond to a constant $\kappa \times 10^5$ value. The values are averages over three wave vectors and error bars (amounting to about 20%) are omitted for clarity

files of the films. The deviation from a purely elastic film increases with increasing side chain length, and is most pronounced for PODcMA (C16). A systematic progression with the side chain length is clear. Thus, it can be claimed that new insights into the monolayer viscoelasticity with respect to the polymer structures are gained from such a polar plot. The trend, however, vanishes altogether once the system moves away from the dilute regime. When the surface mass density is increased to the semi-dilute regime, the data are all crowded around Limit V, obscuring their viscoelastic behaviors. Because of the difficulty in analyzing in the vicinity of Limit V, we turn to an alternative method to analyze the behavior in the semi-dilute regime. Here, we adopt a rather well-established rheological scheme [116]. Instead of just ε_d and κ, we recast the two into the dilational storage modulus and the corresponding loss modulus. Hence, the former is just equal to ε_d while the latter is $\omega_0 \kappa$. Correspondingly, we can define the loss tangent,

$$\tan \delta \equiv \frac{\omega_0 \kappa}{\varepsilon_d} \tag{41}$$

Since we have on hand ε_d and κ as functions of A and Π (Fig. 22), we plot $\tan \delta$ against A and Π in double logarithmic scales in Fig. 23. Now, these provide a clear scheme for the demarcation between the dilute and semi-dilute regimes. In the dilute regime, $\tan \delta$ of each polymer is distinguishable and monotonically decreasing with decreasing A, and the trend follows closely the progressive enhancement of the viscous contribution to ε^* as shown in Fig. 24. A polymer with a longer side chain has a higher $\tan \delta$ value, thus is more viscous than one with a shorter side chain; although the trend is somewhat obscured with PODcMA, we can see that the monolayer with a longer side chain shows greater loss relative to energy storage. Presumably the longer side chains interact more dissipatively with the A/W interface.

In the semi-dilute regime, decreasing A increases $\tan \delta$ for all four PXcMA polymers, producing a limiting scaling behavior of $\tan d \sim A^{-10} \sim \Pi^2$ regardless of the side chain length. This is shown in Fig. 25. The plots make it apparent that $\tan \delta$ shows a clearer scaling with A than with Π at high surface mass densities; a clearer scaling with A is not surprising since A is an independent variable while Π is a dependent one, deduced from the surface tension measurements. The finding that all four collapse onto a single double logarithmic plot makes the following interpretation rather compelling. In the semi-dilute regime, $\tan \delta$ is more closely related to the surface mass density of monomers at the interface A than to the surface pressure Π, which is not only related to the mass density of monomers but also the length of side chains. Thus, the side chains lie down or thermally fluctuate and interact with the interface (Fig. 26a). With increasing lateral packing or decreasing A, the side chains start to leave the interface and the viscous contribution to $\tan \delta$ decreases. In the semi-dilute regime the side chains are standing up in the air and only the backbone can interact with the air/water interface (Fig. 26b).

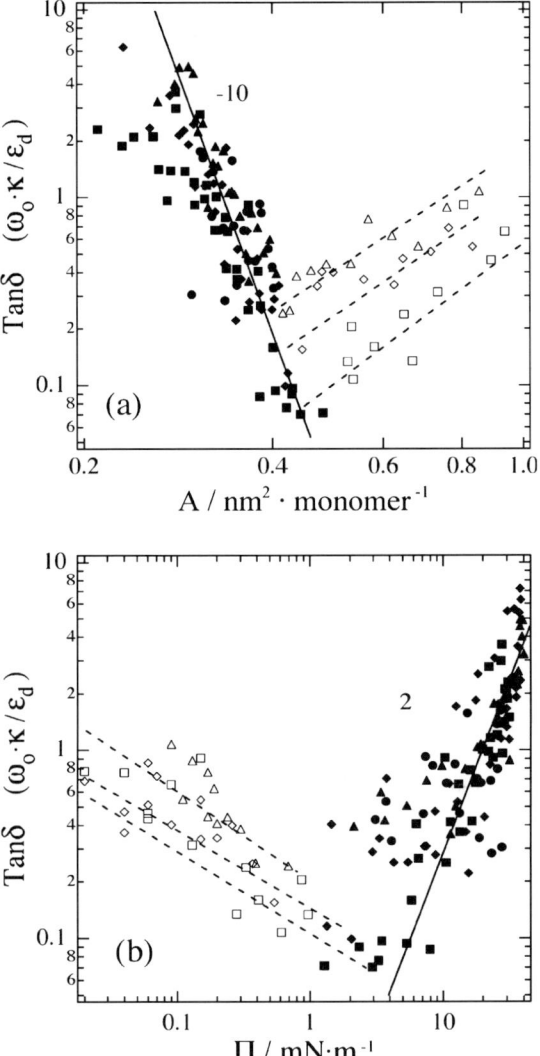

Fig. 25 Loss tangent (tan δ) for PXcMA polymers on 0.01 M HCl at 25 °C. The *symbols* are the same as Fig. 2. The *dashed lines* are guidelines for each polymer and the *solid line* shows the slope of tan δ at high Π or at small A; $\tan\delta \sim A^{-10} \sim \Pi^{2}$. The values represent averages over three wave vectors with error bars amounting to about 30% omitted for clarity. In both plots, PHcMA data in the dilute regime are not included since they are scattered much more than the rest, whereas the conclusion drawn for the semi-dilute regime is not affected by the omission

The results of this study are summarized as follows. The principal mechanism of viscous dissipation of this series of PXcMA polymers at the A/W interface occurs by interactions between the polymer chains and the inter-

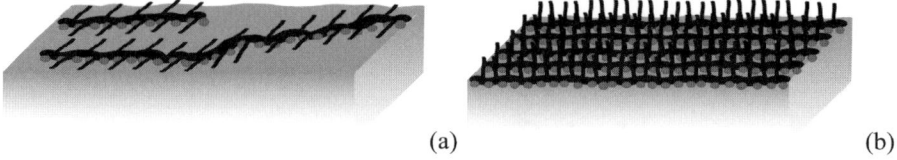

Fig. 26 Conjecture of chain conformation is illustrated for PXcMA polymer monolayers at A/W (**a**) in the dilute regime and (**b**) in the semi-dilute regime. Gray balls represent hydrophilic carboxyl groups and black lines represent hydrocarbon chains

face rather than by interactions between the polymer chains. In the dilute regime, the surface viscoelasticity of the monolayers can be tuned by changing the side chain length of the α-olefin. However, in the semi-dilute regime the surface viscoelasticity is mainly dependent on the surface mass number density, not on the side chain length, and the loss tangent shows a limiting scaling: $\tan\delta \sim A^{-10} \sim \Pi^2$. This distinctive difference in viscoelasticity may be related to the side chain orientation. At low concentrations, the side chains lie on the surface and dissipate energy through interactions with the interface, so a polymer with longer side chains dissipates more energy than one with shorter side chains. The increasing surface concentration in the semi-dilute regime causes the side chains to stand up, so energy dissipation occurs mainly through the interaction between the interface and the polymer.

5.2.2
Block Copolymers

There is a vast body of diblock copolymer studies since block choice can be such that they resemble amphiphilic surfactants. For the sake of brevity, we will skip them. Instead, we present an interesting case of triblock copolymers of poly(ethylene oxide), PEO, and poly(propylene oxide), PPO, commonly known by one of its trade names, Pluronics [117]. They have been used as non-ionic surfactants for a variety of applications such as in emulsification and dispersion stabilization. In aqueous solutions, these copolymers form micelles, and there exists a well-defined critical micelle concentration that is experimentally accessible. Several groups have investigated colloidal suspensions of these polymers [118–122]. The surface properties of the adsorbed monolayers of the copolymers have been reported with respect to their structures and static properties [123–126].

The critical micelle concentration (CMC) of Pluronics [121] has been found to be rather sensitive to temperature, which is ascribed to the changes in hydrophilicity of the ethylene oxide moiety with temperature. As shown already, monolayer viscoelasticity is correlated with the hydrophilicity of polymer backbones such as in polyethers, and that of the pendant group in vinyl polymers. It has been shown that PPO acts differently as a polyether than PEO or PTHF. For this system of triblock copolymers, the PEO segment serves as the hydrophilic component, whereas the PPO segment acts as the hydrophobic segment. Our focus here is to examine whether there exists corresponding temperature sensitivity to the surface viscoelasticity of these copolymers although we can access a very limited range of temperature because of vaporization problems at high temperatures.

The results for two samples of Pluronics, L44 and 10R5, are presented here; sample I (L44) has the structure $(EO)_{10}$-$(PO)_{22}$-$(EO)_{10}$ with an overall molecular weight of 2200 g/mol and sample II (10R5) has the structure $(PO)_8$-$(EO)_{22}$-$(PO)_8$ with the corresponding molecular weight of 1950 g/mol, where EO represents the $-CH_2-CH_2-O-$ unit and PO represents the $-CH_2-CH(CH_3)-O-$ unit. Surface mss density Γ is calculated with the assumption that all the polymers spread on the surface remain there without desorption into the subphase, allowing for the possibility that the extent of the PEO segment submerging into the subphase increases with temperature. Thus, Γ is a more relevant independent variable than $<A>$, average area per repeating unit.

In Fig. 27 are displayed (a) Π-Γ isotherms and (b) the static elasticity ε_s as a function of Γ for sample I (PEO-PPO-PEO) at three different temperatures over the limited range of 9–30 °C. It appears that Π and ε_s, before reaching a maximum $\varepsilon_{s,max}$ at 9 °C, are marginally higher than those at 23 °C and 30 °C. Also, the values of critical surface concentration Γ^* when $\varepsilon_{s,max}$ is reached, seemingly increases with temperature although the same trend does not persist for the interval 23–30 °C. In other words, the trend indicates that the nominal area per unit mass ($A^*_{nominal} = 1/5$) decreases with increasing temperature. In terms of earlier studies with PEO-PS diblock copolymers and PEO homopolymers [127], we can assume that the fully covered monolayer state persists up to Γ^*. If so, the temperature dependence may qualitatively be explained as follows. The amount of PEO block submerging into the subphase increases with increasing temperature because of increasing solubility and/or decreasing surface activity of the PEO block. Alternatively, consistent with the proposal based on the temperature dependence of critical micelle concentration of these triblock copolymers by Alexandridis et al., the observed temperature dependence can arise from less hydration of the PEO block and/or increasing hydrophobicity of the PPO block, i.e. polymer blocks losing water molecules, with increasing temperature due to conformational change of the polymer in the solution with temperature change. A similar argument was delivered by Shuler and Zisman in 1970 [128] that the chain conformation on the air/water interface begins to change from the *trans*-conformation, which has 20 Å^2/monomer as a non-hydrated area per monomeric unit and 28 Å^2/monomer as a hydrated area, to the *cis*-conformation which has 16 Å^2/monomer as a non-hydrated area when the surface film gets compressed to less than 20 Å^2/monomer. But the critical concentration of sample I (L44) at 9 °C, $\Gamma^* = 0.4$ mg/m^2, corresponds to $A^*_{nominal}$

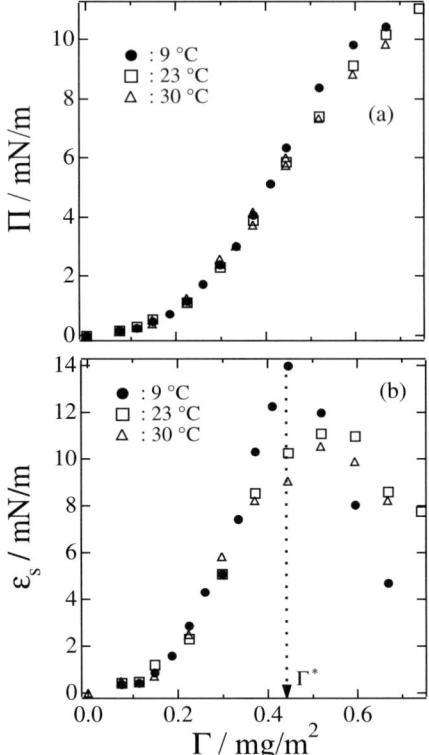

Fig. 27 $\Pi - \Gamma$ isotherms (**a**) and $\varepsilon - \Gamma$ for sample I (PEO-PPO-PEO) on the air/water interface at various temperatures; 9 °C (●), 23 °C, (□) and 30 °C (△). Γ^* indicates the surface mass density where the static elasticity at 9 °C reaches the maximum, not the onset point of semi-dilute solution

of 17 Å^2/monomer and Γ^* at 23 and 30 °C are about 0.5 mg/m^2 corresponding to A^*_{nominal} of 14 Å^2/monomer. The conformational change with changing area per monomer may be possible, but A^*_{nominal} at higher temperature is even smaller than in the non-hydrated area, 20 Å^2/monomer; and recent experimental data [125] suggested that the PEO blocks form brushes under the surface at higher concentration, which is consistent with our previous argument [127] that some of the end-chain PEO blocks penetrate into the subphase.

We turn to the results obtained with sample II, 10R5 (PPO-PEO-PPO). In Fig. 28 are shown the results for this sample: (a) Π-Γ isotherms and (b) the static surface elasticity ε_s as a function of Γ at three different temperatures. Relative to Π-Γ isotherms and the surface concentration Γ^* (equaling 0.4 mg/cm^2) when ε_s reaches $\varepsilon_{s,\text{max}}$, there is no temperature dependence which is in complete contrast to those for sample I; no temperature dependence can be discerned within experimental uncertainties in this case.

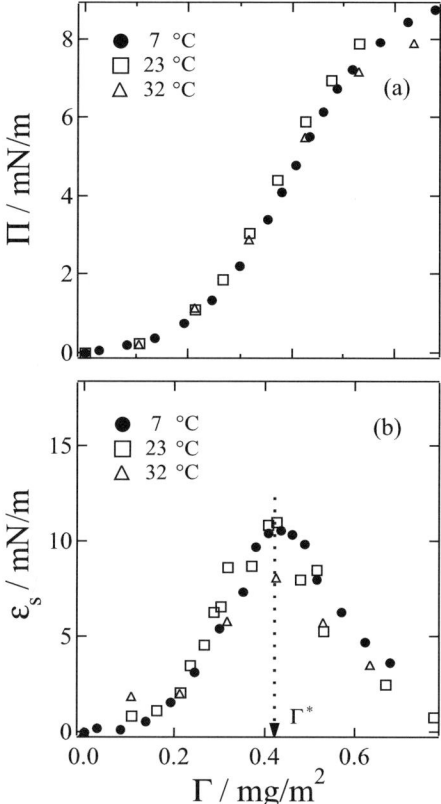

Fig. 28 $\Pi - \Gamma$ isotherms (**a**) and $\varepsilon - \Gamma$ for sample II on the air/water interface at various temperatures; 7 °C (●), 23 °C, (□) and 32 °C (△). Γ^* stands for the same as in Fig. 29

A qualitative interpretation of this contrasting behavior is offered as follows. Unlike with sample I where PEO blocks in both ends of PPO are conformationally free to respond to temperature changes here the PEO block in the middle is conformationally pinned by PPO blocks at both ends by the nature of segment connectivity, PPO-PEO-PPO.

We come to the dynamics of these polymer monolayers. At the risk of repetition, we note again that the polar plot has a unique feature relative to resolution. As the surface viscoelasticity increases, the spacing between grids decreases, resulting in broadening of imprecision for the viscoelastic parameters. Thus, clear specifications of viscoelastic parameters are hard to come by when ε_d and κ exceed $20\,\text{mN}\,\text{m}^{-1}$ and $3 \times 10^{-4}\,\text{mN}\,\text{s}\,\text{m}^{-1}$, respectively. Pluronics copolymers under study here pose an intriguing question by virtue of their architectural difference in the hydrophilic EO versus hydrophobic PO moieties. Sample I (with PEO-PPO-PEO) has a greater amount of hydrophilic PEO blocks which conformationally have greater freedom than that in sample

II (with PPO-PEO-PPO) which has a less hydrophilic PEO block that is also conformationally pinned by PPO blocks.

The polar plots of sample I at 9, 23 and 30 °C are shown in Fig. 29 (a) while those of sample II at 7, 23 and 32 °C are displayed in (b). Overall the behavior of the monolayer shows elasticity-dominance. A marginally lower temperature condition shows higher viscosity.

The viscoelastic behavior of sample I (PEO-PPO-PEO) at 9 °C is somewhat different than those at higher temperatures. At 9 °C, (Fig. 30 (a)●) reaches an asymptotic value (~ 14 mN/m^2) and the corresponding viscosity (Fig. 30 (b) ●) drops rapidly to 0 mN s m^{-1} around the critical surface density (≈ 0.4 mg m^{-2}) where the static elasticity shows the maximum value. Whereas in the cases of 23 and 30 °C (Fig. 31 (a) □ and △), the dynamic dilational elasticity reaches an asymptotic value (~ 15 mN m^{-1}) and the dilational viscosity drops to -1×10^{-5} mN s m^{-1} around $\Gamma^* \approx 0.5$ mg m^{-2} which is higher than that at 9 °C. There is no sharp drop of the dilational viscosities at 23 and 30 °C in contrast to the viscosity behavior at 9 °C; and the maximum viscosity value ($\sim 2 \times 10^{-5}$ mN s m^{-1}) at 23 and 30 °C is lower than that ($\sim 4 \times 10^{-5}$ mN s m^{-1}) at 9 °C.

Figure 31 shows ε_d (a) and κ (b) - Γ plots for sample II, 10R5 (PPO-PEO-PPO). Dynamic dilational elasticities ε_d at different temperatures reach their asymptotes at the same concentration (~ 0.5 mg m^{-2}) but a bit higher than Γ^* (0.4 mg m^{-2}). Corresponding viscosities κ at three different temperatures reach the minima at the same value of Γ^* as the static elasticities of this poly-

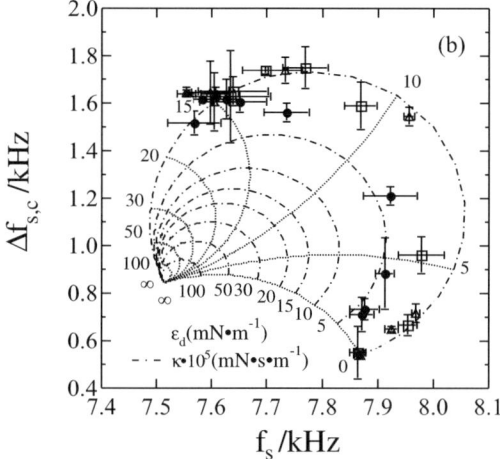

Fig. 29 Polar plot for sample I (**a**) at various temperatures; 9 °C (●), 23 °C, (□) and 30 °C (△) and that for sample II (**b**) at similar temperatures; 7 °C (●), 23 °C, (□) and 32 °C (△). $\Delta f_{s,c,eq}$, and $f_{s,eq}$ are calculated based on the state that $k_{ref} = 324.3$ cm^{-1}, $\eta_{ref} = 0.894$ cP, $\rho_{ref} = 0.997$ g/cm^3, $\sigma_{d,ref} = 71.79$ mN/m, and $\mu_{ref} = 0$ mN s/m. The values represent averages over three wave vectors

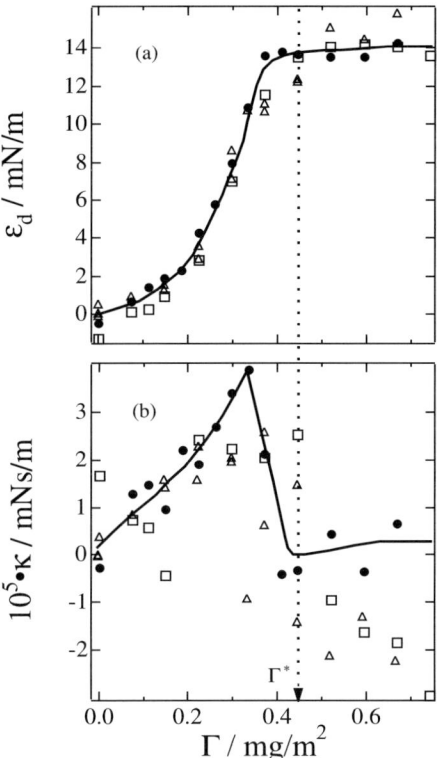

Fig. 30 ε_d and $\kappa - \Gamma$ for sample I on water at various temperatures; 9 °C (●), 23 °C, (□) and 30 °C (△). The viscoelastic parameters represent averages over three wave vectors. *Solid curves* are to guide the eye for the values at 9 °C. Error bars are omitted for clarity. Γ^* comes from Fig. 29b

mer do, while maxima and mass densities are quite different. The maximum dilational viscosities of both polymers decrease with increasing temperature while the values of the maximum dynamic elasticities are almost temperature independent in the range of our observation.

On the basis of our experimental results presented so far, the overall viscoelastic behavior of these triblock copolymers shows an elasticity-dominance over the viscosity. After reaching the critical mass density, where the static elasticity ε_s reaches the maximum, these triblock copolymers collapse into the subphase and form hydrated brushes; and these anchored brushes may be responsible for the result that the surface viscosities drop to around the 0 value at Γ^*. A distinctive difference between two types of polymers, sample I (PEO-PPO-PEO) and sample II (PPO-PEO-PPO), is the temperature dependence of Γ^* where both static elasticity and dilational viscosity show kinds of transitions. Γ^* of sample I increases with increasing temperature while that of sample II does not change with temperature.

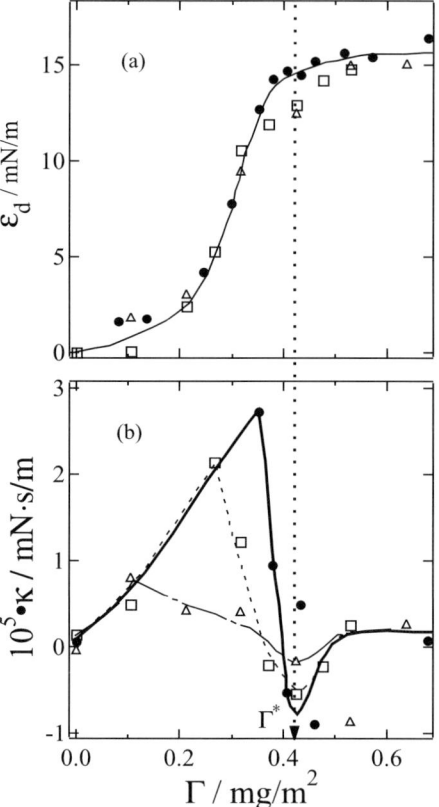

Fig. 31 ε_d and $\kappa - \Gamma$ for sample II on water at various temperatures; 7 °C (●), 23 °C, (□) and 32 °C (△). The viscoelastic parameters represent averages over three wave vectors. *Solid curves* are to guide the eye for the values at 7 °C. The *dotted curve* is for κ at 23 °C and the *line-dotted* curve is at 32 °C. Error bars are omitted for clarity. Γ^* comes from Fig. 29b

6
Other Methods for Monolayer Dynamics

There exists a substantial history of interest in flow and deformation properties of monolayers. Perhaps, the first is the theoretical formulation of hydrodynamic coupling between the monolayer and subphase by Harkins and Kirkwook in 1938 [129], in determination of steady shear viscosity of monolayers, which has since been augmented by Hansen [130] and Goodrich [131]. A variation of the method based on the Maxwell model was proposed by Mannheimher and Schechter [132] in an oscillatory mode in a canal. Experimentally, the method was implemented by joint efforts in our laboratories for determinations of steady shear viscosity of monolayers through the canal viscometry, based on the formulated hydrodynamic coupling [133–136]. Subsequently, Yamamoto's laboratory accomplished the same with another polymer system [137]. The results from these two groups will be discussed below in connection with results obtained by another method. In recent decades, a set of new methods emerged in the study of viscoelasticity of monolayers by means of various mechanical and optical methods.

Among the methods employing externally generated capillary waves, there is one that is uniquely non-invasive by avoiding mechanical contact with the monolayer. It is the technique of electrocapillarity, first extensively explored by Leiderer [138] for electrically charged interfaces of He^3/He^4. For the application to monolayer dynamics on A/W, it was reported by Sohl et al. [139] about the same time. The method is to generate capillary waves by electric field gradient and detect the wave propagation characteristics by means of specular reflection of a laser beam along the propagation direction. The technique was reviewed by Miyano [140] in 1992, with examples of generic surfactants. In our laboratory, we applied the technique [141] to a polymeric monolayer with use of PEO at A/W and O/W.

Turning to various contact methods, they have been extant for some time and some new ones have emerged recently. According to Warburton [142], they may be categorized by means of surface dilatation at various frequencies and by surface shear under steady and oscillatory modes. We list first the methods without regard to their applications to polymer monolayer systems. Some of the conventional methods have been reviewed [131, 143, 144]. As for the newer methods, a device by Kurnaz and Schwarz [145] should be included in the list for the canal method. For the method of the rotating disk and ring, a device reported by Gaub and McConnell [146] consisting of a disk of microscope slide glass driven by a magnetic field through a magnet mounted on the disk is cited first. A somewhat similar device is by Abraham et al. [147], and a further evolution of it by Ghaskadvi et al. [148], using a rotor suspended by a torsion wire, that has a remarkable dynamic range of shear modulus of 10^{-2}–10^3 mN/m. Another method is to drive a magnetic rod longitudinally in an oscillatory mode and is called an Interfacial Stress Rheometer (ISR) by the groups of Fuller and Frank [149], and it has produced some interesting results for polymer monolayers. The last one cited in this list is that of step compression and the oscillatory barrier by Rubio's group [111], which have been directed to probe polymer monolayers under good and theta solvent conditions.

We close this section by examining the status of applications of these methods to polymer monolayers. Initially, ISR was used to probe the 2D nematic state of phthalocyaninatopolysiloxane, descriptively called a "hairy rod", dispersed in eicosanol [149], and subsequently applied to a set of poly(t-butyl methacrylate) in the semi-dilute regime and beyond [150]. In the semi-dilute regime, the surface viscosity is found to scale linearly with molecular weight, which is in good accord with the results of Sacchetti et al. [134]

for the same polymer system. More approximately, they are also in accord with those of Sato et al. [137] for a copolymer system of poly(vinyl octanal acetal) and poly(vinyl alcohol). Most of these methods can indeed extract viscoelastic moduli, i.e., storage and loss moduli. Most extensive in applying a full complement of various methods to span as wide a frequency range as possible are the efforts by the Monroy-Rubio group, exemplified by their efforts with poly(vinyl acetate) and poly(p-hydroxystyrene) in good and poor solvent conditions, respectively. The oscillatory barrier method of this group has undergone some refinement in the data analysis by Fourier transform, and a review appeared recently [151]. A caveat is offered in the examination of these efforts for viscoelastic properties of polymer monolayers with respect to surface mass density. Once the monolayer collapse state has set in as revealed by the surface pressure isotherm, different sorts of non-linear behavior are observed. Such should not be included in the discussion of monolayer properties since the persistence of the monolayer state is rarely proven or offered. Multilayers of polymer chains can scarcely be construed as two-dimensional objects in light of difficulty for strict thermodynamic specification of monolayers to be two-dimensional as we stated in the Introduction.

7
Conclusions

Insoluble polymer monolayers as confined to A/W have received significant attention in the past few decades. While the initial driving force might have been fabrication of Langmuir–Blodgett films, it has now been transformed into a field of polymer dynamics in a reduced dimension, having its own significance in fundamental science. This is further buttressed by the current preoccupation with nanoscience and nanotechnology in the scientific community in the context of a monolayer being a primordial interface object of nanometer thickness. Thus, a review of this sort dealing with the dynamics of 2D objects affords a singular value in the contemporary milieu of science and technology. In the foregoing, we summarized how the capillary wave dynamics of a polymer monolayer covered air/water interface are interpreted to extract their viscoelastic characteristics. The process is to make use of the dispersion equation for the capillary wave propagation under the condition of resonant mode coupling between the transverse and lateral dilational waves. In the final analysis, we describe how elastic and loss components of the dilational modulus are deduced, and their relative proportion depends intimately on the chemical structure of monolayer constituents. Since the static structure of a monolayer as gleaned by the surface pressure isotherm is closely related to the chemical structure of a chain unit, it is entirely predictable that the dynamics probed by the capillary wave propagation should correspond closely to the static structure. Thus, we demonstrate a clean bifurcation of polymer monolayers into good solvent and poor solvent conditions. In so doing, we learn that a corresponding bifurcation into two classes of monolayers, namely the elasticity-dominant and viscosity-dominant polymers is borne out. A down side of this effort, however, is to miss an important attribute of polymers, the chain length dependence. By virtue of having to restrict to a semi-dilute regime of polymer monolayers because of the sensitivity of the SLS technique, it has not been possible to probe the molecular weight dependence of monolayers by the technique. Hence, combining the full complement of methods, as briefly explained in the last section, on top of SLS should give rise to a wider comprehension over the full dynamic range. It appears that such efforts are being initiated.

References

1. Reiter G, Vancso J (2006) Adv Polym Sci 200:XIII
2. Tanford C (1989) Ben Franklin Stilled the Waves: An Informal History of Pouring Oil in Water with Reflections on the Ups and Downs of Scientific Life in General. Duke University Press, Durham, NC
3. Franklin B (1774) Phil Trans 64:455
4. Giles CH (1969) Chem Ind p 1616
5. Giles CH, Forrester SD (1970) Chem Ind p 80
6. Giles CH, Forrester SD (1971) Chem Ind p 43
7. Rayleigh L (1908) Collected Papers, vol I. Cambridge University Press, Cambridge, UK, p 87
8. Thomson W (1871) Philos Mag 42:368
9. Pockels A (1891) Nature 43:437
10. Crisp DJ (1946) J Colloid Sci 1:49
11. Crisp DJ (1946) J Colloid Sci 1:161
12. Gabrielli G, Puggelli M (1971) J Colloid Interface Sci 35:460
13. Petty MC (1996) Langmuir–Blodgett Films. Cambridge University Press, Cambridge, UK
14. Butt H-J, Graf K, Kappl M (2003) Physics and Chemistry of Interfaces. Wiley, Weinheim
15. Ulman A (1991) In: An Introduction to Ultrathin Organic Films: from Langmuir–Blodgett to Self Assembly. Academic Press, San Diego, p 151 and 159
16. Swalen JD (1987) Thin Solid Films 152:151
17. Swalen JD, Allara DL, Andrade JD, Chandross EA, Garoff S, Israelachvili J, McCarthy TJ, Murray R, Pease RF, Rabolt JF, Wynne KJ, Yu H (1987) Langmuir 3:932
18. Adamson AW, Gast A (1995) Physical Chemistry of Surfaces, Chapter IV, 6th ed. Wiley, New York
19. Langmuir I (1917) J Am Chem Soc 39:1848
20. Langmuir I (1936) Science 84:379
21. Takahashi A, Kawaguchi M (1982) Adv Polym Sci 46:1
22. Kawaguchi M, Komatsu S, Matsuzumi M, Takahashi A (1984) J Colloid Interface Sci 102:356
23. Kawaguchi M, Yoshida A, Takahashi A (1983) Macromolecules 16:956

24. Granick S (1985) Macromolecules 18:1597
25. Granick S, Clarson SJ, Formoy TR, Semlyen JA (1985) Polymer 26:925
26. Granick S, Herz J (1985) Macromolecules 18:460
27. Vilanove R, Poupinet D, Rondelez F (1988) Macromolecules 21:2880
28. Vilanove R, Rondelez F (1980) Phys Rev Lett 45:1502
29. Sauer BB, Kawaguchi M, Yu H (1987) Macromolecules 20:2732
30. de Gennes P (1979) Scaling Concepts in Polymer Physics. Cornell University Press, Ithaca, NY
31. Ruberstein M, Colby RH (2003) Polymer Physics. Oxford University Press, New York
32. Le Guillou JC, Zinnjustin J (1977) Phys Rev Lett 39:95
33. Stephen MJ, McCauley JL (1973) Phys Lett A 44:89
34. Takahashi A, Yoshida A, Kawaguchi M (1982) Macromolecules 15:1196
35. Kawaguchi M, Sauer BB, Yu H (1989) Macromolecules 22:1735
36. Braslau A, Pershan PS, Swislow G, Ocko BM, Alsnielsen J (1988) Phys Rev A 38:2457
37. Levich VG (1962) Physicochemical Hydrodynamics. Prentice-Hall, Englewood Cliffs, NJ
38. Katyl RH, Ingard U (1967) Phys Rev Lett 19:64
39. Katyl RH, Ingard U (1968) Phys Rev Lett 20:248
40. Bouchiat MA, Meunier J, Brossel J (1968) CR Acad Sci Ser A B 266B:255
41. Langevin D (1981) J Colloid Interface Sci 80:412
42. Langevin D (1976) J de Phys 37:901
43. Langevin D (1975) J de Phys 36:745
44. Langevin D (1974) J Chem Soc-Faraday Trans I 70:95
45. Langevin D (1972) J de Phys 33:249
46. Langevin D (1992) Light Scattering by Liquid Surfaces and Complementary Techniques. Surfactant Science series 41. Dekker, New York
47. Thomson W (1910) Math Phys Papers 4:76
48. Bouchiat MA, Meunier J (1971) J Phys (Paris) 32:561
49. Dorrenstein R (1951) Koninkl Ned Akad Wetenshap Proc B54:260
50. Goodrich FC (1961) Proc Roy Soc (London) A260:490
51. Goodrich FC (1961) Proc Roy Soc (London) A260:481
52. Goodrich FC (1962) J Phys Chem 66:1858
53. Hansen RS, Mann JA (1964) J Appl Phys 35:152
54. van den Tempel M, van de Riet RP (1965) J Chem Phys 42:2769
55. Lucassen-Reynder E, Lucassen J (1969) Adv Colloid Interface Sci 2:347
56. Kramer L (1971) J Chem Phys 55:2097
57. Lucassen J (1968) Trans Faraday Soc 64:2221
58. Lucassen J (1968) Trans Faraday Soc 64:2230
59. Lucassen J, Hansen RS (1966) J Colloid Interface Sci 22:32
60. Lucassen J, Hansen RS (1967) J Colloid Interface Sci 23:319
61. Lucassen J, Vandente M (1972) J Colloid Interface Sci 41:491
62. Mann JA (1985) Langmuir 1:10
63. Hard S, Neuman RD (1987) J Colloid Interface Sci 120:15
64. Stokes GG (1845) Cambridge Trans 8:287
65. Reynolds O (1880) Brit Ass Rept (cited in Lucassen-Reynders and Lucassen, 1969)
66. Langevin D (1992) Surfact Sci Ser 41:61
67. Earnshaw JC, McGivern RC, McLaughlin AC, Winch PJ (1990) Langmuir 6:649
68. Henderson JA, Richards RW, Penfold J, Thomas RK (1993) Macromolecules 26:65
69. McGivern RC, Earnshaw JC (1989) Langmuir 5:545
70. Tyndall J (1869) Proc Roy Soc London 17:223

71. Fabelinskii I (1968) Molecular Scattering of Light. Plenum, New York
72. Smoluchowski M (1908) Ann Physik 25:225
73. Mandelstam L (1913) Ann Physik 41:609
74. Gans R (1926) Ann Physik 79:204
75. Andronov AA, Leontovich MA (1926) Z Phys 38:485
76. Raman CV, Ramdas LA (1925) Proc Roy Soc A 109:150
77. Mountain RD (1966) Rev Mod Phys 38:205
78. Chu B (1974) Laser Light Scattering. Academic Press, New York
79. Berne BJ, Pecora R (1976) Dynamic Light Scattering. Wiley, New York
80. Bouchiat MA, Meunier J (1968) CR Acad Sci Paris Ser A B 266B:301
81. Bouchiat MA, Meunier J (1969) Polarization, Matter and Radiation. Presses Universitaires, Paris
82. Langevin D (1992) Historical Development. In: Langevin D (ed) Light Scattering by Liquid Surfaces and Complimentary Techniques, Chap 1. Dekker, New York
83. Egelstaff PA (1967) An introduction to the liquid state, Chap 14. Academic, New York
84. Sano M, Kawaguchi M, Chen YL, Skarlupka RJ, Chang T, Zografi G, Yu H (1986) Rev Sci Instrum 57:1158
85. Sano M (1987) PhD Thesis, University of Wisconsin-Madison
86. Sauer BB (1987) PhD Thesis, University of Wisconsin-Madison
87. Lee W (1994) PhD Thesis, University of Wisconsin-Madison
88. McQueen D, Lundstrom I (1973) J Chem Soc Faraday Trans I 69:694
89. Langevin D (1992) Theoretical Background—Scattered Intensity. In: Langevin D (ed) Light Scattering by Liquid Surfaces and Complementary Techniques, Chap 3. Dekker, New York
90. Hard S, Hamnerius Y, Nilsson O (1976) J Appl Phys 47:2433
91. Earnshaw JC, McGivern RC (1988) J Colloid Interface Sci 123:36
92. Esker AR, Zhang LH, Sauer BB, Lee W, Yu H (2000) Colloids Surf A 171:131
93. Ries HE Jr, Walker DC (1961) J Colloid Sci 16:361
94. Thomas RK (2004) Ann Rev Phys Chem 55:391
95. Grundy MJ, Richardson RM, Roser SJ, Penfold J, Ward RC (1988) Thin Solid Films 159:43
96. Richardson RM, Roser SJ (1991) Langmuir 7:1458
97. Lee LT, Mann EK, Langevin D, Farnoux B (1991) Langmuir 7:3076
98. Henderson JA, Richards RW, Penfold J, Shackleton C, Thomas RK (1991) Polymer 32:3284
99. Henderson JA, Richards RW, Penfold J, Thomas RK, Lu JR (1993) Macromolecules 26:4591
100. Gissing SK, Richards RW, Rochford BR (1994) Colloids Surf A 86:171
101. Crisp DJ (1947) Res (London) Suppl Surface Chem 17:23
102. Crisp DJ (1958) In: Danielli JF, Pankhurst KGA, Riddiford AC (eds) Surface Phenomena in Chemistry and Biology. Pergamon Press, London, p 23
103. Langevin D, Meunier J, Chatenay D (1984) In: Mittal KL, Lineman B (eds) Surfactants in Solution. Plenum Press, New York
104. Kawaguchi M, Sano M, Chen YL, Zografi G, Yu H (1986) Macromolecules 19:2606
105. Yoo KH, Yu H (1989) Macromolecules 22:4019
106. Runge FE, Yu H (1993) Langmuir 9:3191
107. Lee WK, Esker AR, Yu H (1995) Colloids Surf A 102:191
108. Chen YL, Kawaguchi M, Yu H, Zografi G (1987) Langmuir 3:31
109. Esker AR (1986) PhD Thesis, University of Wisconsin-Madison
110. Rivillon S, Monroy F, Ortega F, Rubio RG (2002) Eur Phys J E 9:375

111. Monroy F, Rivillon S, Ortega F, Rubio RG (2001) J Chem Phys 115:530
112. Mann JA, Du G (1971) J Colloid Interface Sci 37:2
113. Monroy F, Ortega F, Rubio RG (1998) Phys Rev E 58:7629
114. Monroy F, Ortega F, Rubio RG (2000) Eur Phys J B 13:745
115. Kim C, Esker AR, Runge FE, Yu H (2006) Macromolecules 39:4889
116. Ferry JD (1980) Viscoelastic Properties of Polymers, 3rd ed. Wiley, New York
117. Kim C, Yu H (2003) Langmuir 19:4460
118. Prasad KN, Luong TT, Florence AT, Paris J, Vaution C, Seiller M, Puisieux F (1979) J Colloid Interface Sci 69:225
119. Chu B (1995) Langmuir 11:414
120. Alexandridis P, Athanassiou V, Fukuda S, Hatton TA (1994) Langmuir 10:2604
121. Alexandridis P, Holzwarth JF, Hatton TA (1994) Macromolecules 27:2414
122. Zhou ZK, Chu B (1988) J Colloid Interface Sci 126:171
123. An SW, Su TJ, Thomas RK, Baines FL, Billingham NC, Armes SP, Penfold J (1998) J Phys Chem B 102:387
124. Munoz MG, Monroy F, Ortega F, Rubio RG, Langevin D (2000) Langmuir 16:1094
125. Munoz MG, Monroy F, Ortega F, Rubio RG, Langevin D (2000) Langmuir 16:1083
126. Caseli L, Nobre TM, Silva DAK, Loh W, Zaniquelli MED (2001) Colloid Surf B 22:309
127. Sauer BB, Yu H, Tien CF, Hager DF (1987) Macromolecules 20:393
128. Shuler RL, Zisman WA (1970) J Phys Chem 74:1523
129. Harkins WD, Kirkwood JG (1938) J Chem Phys 6:53
130. Hansen RS (1959) J Phys Chem 63:637
131. Goodrich FC (1973) In: Danielli JF, Rosenberg MD, Cadenhead DA (eds) Progress in Suface and Membrane Science. Academic Press, New York, p 151
132. Mannheimer RJ, Schechter RS (1970) J Colloid Interface Sci 32:225
133. Sacchetti M, Yu H, Zografi G (1993) J Chem Phys 99:563
134. Sacchetti M, Yu H, Zografi G (1993) Langmuir 9:2168
135. Sacchetti M, Yu H, Zografi G (1993) Rev Sci Instrum 64:1941
136. Sacchetti M (1992) PhD Thesis, University of Wisconsin-Madison
137. Sato N, Ito S, Yamamoto M (1998) Macromolecules 31:2673
138. Leiderer P (1979) Phys Rev B 20:4511
139. Sohl CH, Miyano K, Ketterson JB (1978) Rev Sci Instrum 49:1464
140. Miyano K (1992) Externally Excited Surface Waves In: Langevin D (ed) Light Scattering by Liquid Surfaces and Complementary Techniques, Chap 16. Dekker, New York
141. Ito K, Sauer BB, Skarlupka RJ, Sano M, Yu H (1990) Langmuir 6:1379
142. Warburton B (1996) Curr Opinion Colloid Int Sci 1:481
143. Joly M (1972) In: Matejevic E (ed) Surf Colloid Sci. Wiley, New York, p 1
144. Edwards DA, Brenner H, Wasan DT (1991) Interfacial Transport Processes and Rheology, Chap 7. Butterworth-Heinemann, Boston
145. Kurnaz ML, Schwartz DK (1997) Phys Rev E 56:3378
146. Gaub HE, McConnell HM (1986) J Phys Chem 90:6830
147. Abraham BM, Miyano K, Xu SQ, Ketterson JB (1983) Rev Sci Instrum 54:213
148. Ghaskadvi RS, Ketterson JB, MacDonald RC, Dutta P (1997) Rev Sci Instrum 68:1792
149. Brooks CF, Fuller GG, Frank CW, Robertson CR (1999) Langmuir 15:2450
150. Gavranovic GT, Deutsch JM, Fuller GG (2005) Macromolecules 38:6672
151. Hilles HM, Monroy F, Bonales LJ, Ortega F, Rubio RG (2006) Adv Colloid Interface Sci 122:67

Editor: Kwang-Sup Lee

Silicone-Based Polymer Blends: An Overview of the Materials and Processes

Patrice Lucas · Jean-Jacques Robin (✉)

Institut Charles Gerhardt Montpellier, UMR 5253 CNRS-ENSCM-UM2-UM1,
Equipe: Ingénierie et Architectures Macromoléculaires,
Université Montpellier II - Bat. 17 – CC1702, Place Eugène Bataillon,
34095 Montpellier cedex 5, France
Jean-Jacques.Robin@univ-montp2.fr

1	Introduction	113
2	Silicone/Polymer Bicomponent Blends	114
2.1	Unmodified Silicone/Polymer Blends	114
2.2	Functionalized Silicone/Polymer Blends	117
3	Compatibilization of Polymer Blends Using Copolymers	118
3.1	Preformed Copolymer Addition	119
3.2	In-Situ Copolymer Formation	122
3.2.1	In-Situ Block Copolymer Formation	123
3.2.2	In-Situ Graft Copolymer Formation	124
3.2.3	In-Situ Branched Copolymer Formation	126
4	Interpenetrating Polymer Networks	128
4.1	Silicone Crosslinking Methods	129
4.2	IPNs Synthesis	130
4.2.1	Sequential IPNs	130
4.2.2	Simultaneous IPNs	134
4.2.3	Latex IPNs and Latex Semi-IPNs	135
4.2.4	Semi-IPNs	136
4.2.5	Mixing Method	138
5	Conclusion	140
	References	140

Abstract Although extensive studies on copolymers have been carried out with a view to exploiting the combined homopolymer properties, physical blends of polymers have warranted less attention. But as a result of increased scientific and economic interest research in this challenging field has grown over the last two decades. The unique properties of silicone polymers, due to their Si – O – Si backbone, including their low Tg's, gives rise to some specific applications. However, it is their singular structure which also makes silicone polymers incompatible with most other macromolecules and limits their incorporation to low amounts. Bleeding and mechanical loss are observed at higher percentages. This overview is divided into three parts: the first covers silicone/polymer bicomponent blends with the silicone being either functionalized or not. The second part describes the different ways to compatibilize the two phases of the silicon/polymer blend using copolymers which can be added as either preformed copolymers or synthesized in situ. The efficiency of the copolymers involved varies depending on their chemical structure and architecture. The final section is dedicated to the different methods of preparation of Interpenetrating Polymer Networks (IPNs) which are commercially and industrially by far the most interesting. The relevant processes (extrusion, batch, casting, etc.) as well as the properties of the various resulting materials are also reviewed throughout the paper.

Keywords Compatibilization · IPN · Silicone/polymer blend

Abbreviations

γ	interfacial tension
λ	viscosity ratio
AFM	atomic force microscopy
AIBN	2,2′-azobis(isobutyronitrile)
BuMA	butyl methacrylate
Ca	capillary number
DCP	dicumyl peroxide
DMA	dynamic mechanical analysis
DMS	dynamic mechanical spectroscopy
DSC	differential scanning calorimetry
EGDMA	ethylene glycol dimethacrylate
EMA	ethyl methacrylate
EPDM	ethylene-propylene-diene rubber
FT-IR	Fourier transform-infra-red
HDPE	high density polyethylene
HTV	high temperature vulcanization
IPN	interpenetrating polymer network
LDPE	low density polyethylene
LLDPE	linear low density polyethylene
MA	maleic anhydride
MAA	methacrylic acid
MDI	4,4′-diphenylmethanediisocyanate
MMA	methyl methacrylate
PA	poly(amide)
PAC	poly(acrylate)
PB	poly(butadiene)
PBT	poly(butylene terephtalate)
PBuMA	poly(butyl methacrylate)
PDMS	poly(dimethylsiloxane)
PDMS-NH$_2$	amino-terminated poly(dimethylsiloxane)
PDMS-COOH	carboxy-terminated poly(dimethylsiloxane)
PE	poly(ethylene)
PEMA	poly(ethyl methacrylate)
PEO	poly(ethylene oxide)
PIP	poly(isoprene)
PMAA	poly(methacrylic acid)
PMMA	poly(methylmethacrylate)
PMPS	poly(methylphenylsiloxane)

PP	poly(propylene)
PPO	poly(propylene oxide)
PS	poly(styrene)
PTFE	poly(tetrafluoroethylene)
PTMO	poly(tetramethylene oxide)
PU	poly(urethane)
PVP	poly(vinylpyridine)
RTV	room temperature vulcanization
SEC	size exclusion chromatography
SEM	scanning electron microscopy
STPV	silicon thermoplastic vulcanizate
TEM	transmission electron microscopy
TEOS	tetraethoxysilane
TPE	thermoplastic elastomer
TPU	thermoplastic poly(urethane)
TPV	thermoplastic vulcanizate
VTES	vinyltriethoxysilane,
XPS	X-ray photoelectron spectroscopy

1
Introduction

The great variety of available monomers combined with extensive research studies in the field of copolymerization allow us to access many advanced technical polymers and copolymers. Hence, it is possible to combine the properties of different homopolymers either to combine the different characteristics or to overcome a particular drawback of one of the homopolymer components. The blending of homopolymers is another rapidly growing technique with much potential. Its main advantage lies in the fact it employs simple processes that are attractive economically and are less time-consuming. The blending of polymers involves some quite different concepts than those in copolymerization. Whereas covalent bonds hold together copolymer components, the homopolymers of a blend are held together by chemical and/or physical interactions. Since for most common polymers these are weak Van der Waals forces, the miscibility is usually not thermodynamically favourable. This is especially true in the case of silicone polymers since the constitutive $(OSi(Me)_2)$ units do not permit the establishment of strong interactions with the other polymer. A widespread method used to decrease the interfacial tension consists of the use of copolymers as compatibilizers. These can be either added as a preformed copolymer or synthesized in-situ. The synthesis of such copolymers is also reviewed here because of its distinctive chemistry in the silicone field. Besides these thermodynamical considerations, the rheological properties of both the homopolymers and the compatibilizers as well as the mixing process play an important role in the development of the final blend morphology and consequently its mechanical properties. Several theoretical and experimental studies have appeared which deal with these complex issues in an attempt to make it possible to identify some universal rules. Results obtained in the context of silicone/polymer blends which are reviewed here are compared as far as possible to the general principles of polymer blends. The reason we focus on silicone blends is because of their potential in a wide range of applications thanks to their very specific properties. The main characteristics of polydimethylsiloxanes include for example their low glass transition temperatures, heat stability, waterproof properties, resistance to oxidation, stability at high and low temperatures, great molecular flexibility, high impact resistance, good electrical insulation, resilience, oxygen permeability, biocompatibility, low surface energy and relative insensitivity to UV light etc.

As a result of these important properties, different blends have been studied: thermoplastics or thermosets with silicone, hydrogel/silicones. These materials can be more or less efficiently produced thanks to the choice of routes: simple and compatibilized blends, as well as IPNs.

2
Silicone/Polymer Bicomponent Blends

2.1
Unmodified Silicone/Polymer Blends

Polysiloxane is well known to be immiscible with most polymers. The resulting materials usually present a two-phase morphology. A good dispersion can be obtained when a special mixing procedure is applied to the blend but it still appears to be unstable. Hence literature describing simple blends of a silicone with another polymer using the usual processes such as extruder or batch mixing is relatively poor. Historically, some patents claimed blends of polysiloxane with polycarbonate [1] which presented good elongation, flexibility and resistance to basic hydrolysis. No exudation of the polysiloxane was observed. The major drawback observed was the difficulty in incorporating the silicone into the thermoplastic. Bostick et al. [2] claimed to avoid this problem by using octaphenylcyclotetrasiloxane instead of a polysiloxane, but the blend properties were not specified. Another example dealt with the incorporation of polydimethylsiloxane into polyamide [3]. Although a better impact strength may have been measured on the freshly extruded material with 6 wt %, the homogeneity is not discussed in the patent and no indication concerning the evolution of the blend is given. Except for these limited examples, exudation and loss of mechanical properties are usually observed. So, in general, the amount of silicone incorporated is limited and should not exceed 2 wt % [4–6]. Only a few percent of polysiloxane is needed to achieve wear-resistant properties and therefore it was used with polyurethane [4, 5] to lower the friction coefficient. An optimal amount of added silicone (1.5 to 2 wt %) increased wear-resistance by up to 25%. Moreover, the ultimate tensile strength and elongation at break were also both enhanced. The thermoplastic poly(urethanes) (TPU) involved in the system (Fig. 1) were composed of both soft segments (poly(tetramethylene oxide): PTMO) and hard segments (4,4′-diphenylmethanediisocyanate: MDI, extended with 1,4-butanediol). The phase separation between the two kinds of segment is responsible for the good mechanical properties of the TPU. However, it was found that the incorporation of PDMS had no effect. The PDMS is incorporated in the soft domains because it is more miscible with PTMO than with TPU hard segments. However, the authors showed that the soft segment volume did not increase whereas the density did. This is a clear demonstration of the chain packing effect of the soft segments by the PDMS and explains the enhanced Young's modulus and tensile strength. Last but not least, the hydrophobicity was also increased.

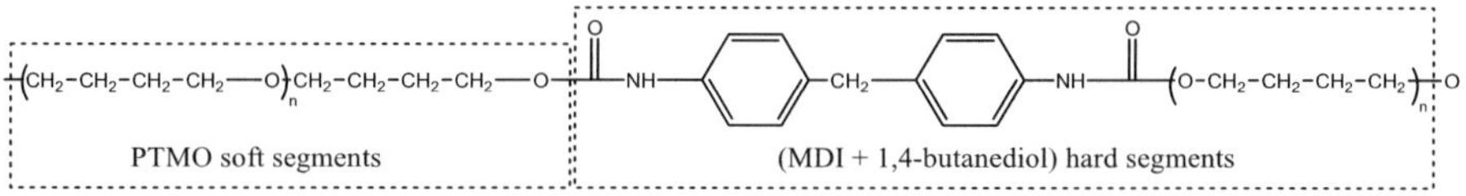

Fig. 1 TPU used by Hill et al. [4]

It should be noted that this type of blend (PU/PDMS) presents an improved chemical resistance [6] that makes them more base-resistant than acid-resistant. In addition, the relative viscosity η_r (or λ) which is defined as the ratio of the dispersed phase viscosity (η_d) to the matrix viscosity (η_m) plays an important role in the quality of the blend morphology formation. An empirical relationship [7] links the capillary number (Ca) to the viscosity ratio (Eq. 1).

$$Ca = 4(\eta_d/\eta_m)^k \qquad k = 0.84 \text{ for } \lambda \geq 1 \text{ and } k = -0.84 \text{ for } \lambda \leq 1. \qquad (1)$$

Since the capillary number is related to the drop deformation under shear stress during mixing and drop break up occurs for a maximum critical value of Ca (Ca$_{cr}$), a η_r value close to unity is required in order to optimize the structure of the blend. To the best of our knowledge, no such study involving polydimethylsiloxane discretely dispersed in a matrix is dedicated to the viscosity ratio influence in binary blends. Nonetheless, Maric and Macosko [8], in a work on copolymer effect that will be discussed later, measured the evolution of the morphology with the viscosity ratio ($\langle D \rangle$ vs. λ) of a PS/PDMS blend (Fig. 2). Their results fall in agreement with theory but, as expected, these blends were not thermodynamically stable and coarsening was observed after annealing.

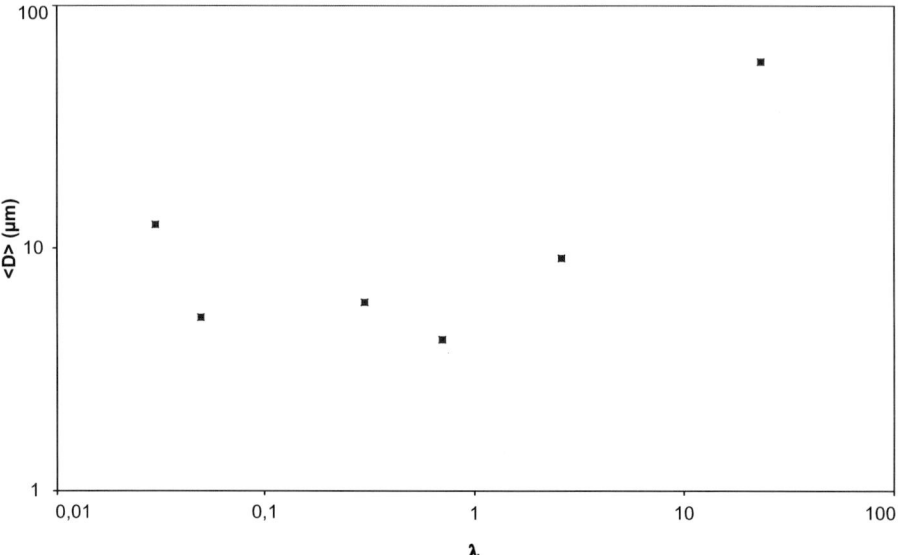

Fig. 2 Viscosity ratio λ versus PDMS drop size ($\langle D \rangle$) for PS/PDMS blends [8]

Chuai et al. [9] studied the influence of the viscosity ratio on the morphology of a PS/PDMS blend in order to determine the composition range where a co-continuous structure is obtained. Since a good synergy of the homopolymer properties arises from co-continuous blends, they are of great interest. For more information, the reader is referred to the excellent review by Pötschke and Paul [10]. Chuai et al. compared three different PS/PDMS systems with varying PS viscosities and fixing the PDMS one. Using a typical extraction method, they measured the degree of continuity (Eq. 2) for each pair of homopolymers for various polymer compositions.

$$\Phi_{i,1} = \frac{\text{mass of polymer 1 (residual or extracted)}}{\text{mass of polymer 1 in the sample before extraction}}. \quad (2)$$

They observed that the range of compositions was broader for the PS/PDMS pair having the smaller viscosity ratio. This tendency is in good agreement with the predicted relationship [11–14]:

$$\frac{\eta_1(\dot{\gamma})}{\eta_2(\dot{\gamma})} \approx \frac{\Phi_{1,\text{PI}}}{\Phi_{2,\text{PI}}} \quad (3)$$

$$\Phi_{2,\text{PI}} = \frac{1}{(1+\lambda)}. \quad (4)$$

In Eq. 3, 1 and 2 are the blend components, η the viscosity and $\dot{\gamma}$ the shear rate applied for mixing the blend. Equation 4 is obtained from Eq. 3 with $\lambda = \eta_1/\eta_2$ and $\Phi_{2,\text{PI}} = 1 - \Phi_{1,\text{PI}}$. However, accurate calculated theoretical values were systematically found to be above the experimental values. This could be explained by the fact that the λ values of the PS/PDMS pair were far from unity considering that Eqs. 3 and 4 give better results for equiviscous polymer blends. But, the Utracki model [15] (Eq. 5) also gave overestimated values although it is even more dedicated to blends having viscosity ratios far from unity. Chuai et al. suggest that a model taking elastic behaviour into account could give better results.

$$\lambda = \left[\frac{(\Phi_m - \Phi_{2,\text{PI}})}{(\Phi_m - \Phi_{1,\text{PI}})} \right]^{[\eta]\Phi_m} . \tag{5}$$

In Eq. 5, $[\eta]$ is the intrinsic viscosity of the dispersed phase and Φ_m is the maximum packing volume fraction (in most cases, $\Phi_m = 1 - \Phi_{cr}$, Φ_{cr} is the critical volume fraction or percolation threshold).

The morphology is also dependant on thermodynamic factors such as the interfacial tension (γ): in the case of a polymer dispersed in another one, the size of the droplets was found to be proportional to the interfacial tension. Furthermore, the interfacial tension gives direct information on the equilibrium state of the blend since it is linked to the free energy (ΔG) of the blend via the Flory–Huggins interaction parameter. Anastasiadis et al. [16] developed the use of an automated pendant drop apparatus coupled with a shape analysis algorithm to obtain the γ values for polymer blends, in particular for a polybutadiene(PB)-polydimethylsiloxane system. The interfacial tension was found to decrease slightly linearly with temperature. It was also found to increase with the PDMS molecular weight (keeping the PB molecular weight constant) in relative good agreement with a $Mn^{-1/2}$ expression instead of the $Mn^{-2/3}$ expressed by other researchers with simpler systems (e.g. n-alkanes vs. low molecular weight PDMS) [17, 18]. The deviation was attributed to the polydispersity of the PDMS. Standard D_4 (octamethylcyclotetrasiloxane) polymerization usually leads to silicone polymers having a polydispersity index around 2. The other explanation was the proximity of the critical point where a $Mn^{-1/2}$ relationship is expected.

2.2
Functionalized Silicone/Polymer Blends

Functionalized polysiloxanes are attractive because, with only small modification of the polysiloxane properties (e.g. density, yield strength, etc.), they allow reduction in the interfacial tension thanks to better interactions with the other homopolymer. Two studies involving PB/PDMS and polyisoprene (PIP)/PDMS are of particular interest [19, 20]. The PDMS end groups were either amine ($-NH_2$) or acid ($-COOH$). It was first observed that the PB/PDMS-NH_2 system exhibits a 30% reduction in interfacial tension compared to the equivalent PB/PDMS system. A preliminary reduction in interfacial tension was also observed for polymers bearing side groups in systems such as PDMS(-g)-NH$_2$/PEO, PDMS(-g)-COOH/PEO [21] and PDMS(-g)-NH$_2$/PB [22]. In a PB/PDMS-NH$_2$ system, a decrease in the interfacial tension as a function of Mn^{-1} was observed but not the expected Mn$^{-2/3}$ relationship generally encountered for surfactant activity. Moreover, the addition of NH$_2$-terminated PDMS in a PB/PDMS system showed a linear relationship between the interfacial tension and the proportion of added PDMS-NH$_2$ (up to 75%). These phenomena led the authors to conclude that the end-group effect is a bulk effect and not an interfacial one. Their conclusion is based on the fact that a surfactant does not involve such linear relationships and only a few percent is required to observe interfacial activity. Interaction parameters deduced from theory or obtained from cloud point curves allowed the authors to suggest that the observed compatibilization is more likely due to repulsion between the PDMS backbone and its amine end groups rather than to unexpected attractive interactions between the amine groups and the PB. The addition of functional PDMSs such as PDMS-NH$_2$, PDMS-g-NH$_2$, PDMS-COOH or PDMS-g-COOH in a PA/PDMS blend has been patented [23]. For example, the addition of 1 wt % of carboxy-terminated polysiloxane to a 75/25 (wt.%) PA/PDMS system reduced the diameter of the PDMS phase from 35 µm to 1–2 µm (SEM measurements). The inventors observed that the diameters of the dispersed particles in the polysiloxane phase were much smaller in the presence of the additive than without it. Moreover, the mechanical properties [tensile elongation and Izod strength (impact resistance)] were enhanced. Finally, Li et al. [24] compatibilized PDMS and polyvinylpyridine (PVP) via hydrogen bonding interactions that they obtained by functionalization of the PDMS with carboxylic acid groups. Using FT-IR, DSC and XPS measurements, the authors concluded that a minimum of 23 mol % was required to obtain a compatibilized blend.

3
Compatibilization of Polymer Blends Using Copolymers

The use of copolymers as surfactants is widespread in macromolecular chemistry in order to compatibilize immiscible blends. These additives are sometimes named "surfactants", "interfacial agents" or more usually "compatibilizers". Their effect on improving different properties is observed: interfacial tension and domain size decrease, while there is an increase in adhesion between the two phases and a post-mixing morphology stabilization (coalescence prevention). The aim of the addition of such copolymers is to obtain thermodynamically stable blends, but the influence of kinetic parameters has to be kept in mind as long as they have to be mastered to reach the equilibrium. Introducing a copolymer can be achieved either by addition of a pre-synthesized copolymer or by in-situ surfactant synthesis via a fitted reaction (reactive blending). The literature focused on the block copolymers based on silicone synthesis is wide [25, 26]. Here, the main drawback of using tailored copolymers is that they can form micelles instead of stabilizing the interface [27]: they have to be long enough to produce an efficient entanglement with the homopolymers but not too long to diffuse quickly to the interface. This compromise is quite difficult to reach. Moreover, in comparison with in-situ synthesized copolymers, they are more expensive which may make them less attractive commercially. Hence, currently this field of research remains essentially of academic interest.

3.1
Preformed Copolymer Addition

A three-component system can consist of A/B/A-B, A/B/C-B, A/B/C-D or A/B/A-B-A where A and B are the homopolymers, C and D are the moieties miscible with A and B, respectively. The influence of many factors have been experimentally studied: copolymer concentration, copolymer and polysiloxane molecular weights, mixing technique, relative viscosity, chemical composition of the copolymer and molecular architectures. The compatibility may be estimated using several representative measurements. The main ones are (i) the interfacial tension (γ) measurements (pendant drop, sessile drop and more recently, dynamic interfacial tension measurements [28, 29]; the latter being more appropriate for polymers presenting high molecular weight and high viscosities) which reflect the thermodynamic stability of the blend. As already mentioned, compatibilization induces a decrease of the interfacial tension; (ii) the microscopy techniques [transmission electron microscopy (TEM), scanning electron microscopy (SEM), etc.] enable researchers to evaluate the different domain size variations and (iii) calorimetric measurements (glass transition and melting point shifts).

Since they act as surfactants, copolymers are added in only small amounts, typically from a thousandth parts to a few hundredth parts. Theoretically, Leibler [30] showed that only 2% of a diblock copolymer may thermodynamically stabilize an 80%/20% incompatible blend with an optimum morphology (submicronic droplets). However, in practice kinetic control and micelle formation interfere in this best-case scenario. To a some extent, compatibilization increases with copolymer concentration [8, 31, 32]. Beyond a critical concentration (critical micellar concentration: cmc) little or no improvement is observed (moreover, for high amounts, the copolymer can act as a plasticizer). Copolymer molecular weight influence is similar to that of the concentration effect. For example, in a PS/PDMS system [8, 31, 32], when the copolymer molecular weight increases, domain size decreases to a certain extent. Hu et al. [31] correlated their experimental results with theoretical prediction of the Leibler's brush theory [30]. Leibler distinguishes two regimes to characterize the behaviour of the copolymer at the interface between homopolymers and consequently the interfacial tension reduction decrease. These two regimes are defined by three parameters: area per copolymers chain (Σ), the block lengths of the copolymer (N_i) and the homopolymers polymerization degrees (P_i). On the one hand, the "wet brush" regime is predicted for $N_i > P_i^{3/2}$ and a relatively large Σ. In this case, homopolymer chains can penetrate the brushes. On the other hand, the "dry brush" regime is predicted for $N_i < P_i^{3/2}$ and a small Σ. Here a lamellar interfacial layer made of copolymer is expected. As an example, for a symmetric system (A/B/A-B, $N_A = N_B$, $P_A = P_B$), Leibler predicts respectively for the "wet brush" and "dry brush" regime, a decrease of the interfacial tension that follows Eqs. 6 and 7:

$$\Delta\gamma \propto 0.12(\ln\varphi + 0.5\chi N)^{5/2} N^{-3/2} P \tag{6}$$

$$\Delta\gamma \propto (2^{3/2}/9)(\ln\varphi + 0.5\chi N)^{3/2} N^{-1/2}. \tag{7}$$

ϕ is the copolymer concentration and χ is the Flory interaction parameter. As can be deduced from these equations, the efficiency of the copolymer should be proportional to the copolymer molecular weight in both cases. The interfacial tension reduction is dependant on the homopolymer molecular weight only in the "wet brush" regime. The interfacial tension decreases with the copolymer concentration until the cmc is reached. Leibler noticed, however, that the formation of micelles can go with the saturation of the interface, hence limiting the interfacial tension reduction. Experimentally, Hu et al. [31] varied the molecular weight of the PDMS in PS/PDMS/PS-b-PDMS blends. In a first blend, the "dry brush" regime was expected thanks to the respect of $P > N^{2/3}$. Their results in the decrease of the interfacial tension were in very good agreement with the theory. Their second blend fell into the crossover region between dry and wet brush, so that experimental tension reduction was lower than the predicted one. It seemed that in addition to the interfacial localization of the copolymer, formation of micelles occurred.

With respect to stability after annealing, an optimal molecular weight effect is observed. Below this molecular weight value, the limited length of the copolymer fails to prevent the blend from coalescing. Higher molecular weight copolymer inhibits efficient diffusion to the interface. For the PS/PDMS/PS-co-PDMS system with respect to the homopolysiloxane molecular weight, its increase leads to a decrease in interfacial tension [31]. When the evolution of the morphology is studied using SEM, it is seen that an optimum PDMS molecular weight is required to reach its ideal dispersion in the matrix [8]. A relative viscosity η_r close to one—just as in the case of simple blends—was also found to give better dispersion [8]. The mixing procedure is also important. Hu et al. [31] observed that in the case of a PS/PDMS/PS-co-PDMS system, the best results were obtained when the copolymer was introduced first into the PDMS component rather than into the PS homopolymer. The authors argued an easier micelle formation in the PS phase keeping in mind that the critical micelle concentration (cmc) of the copolymer is very different depending on the homopolymer in which it is first dissolved. In this case, the ϕ_{cmc} is ~ 1% in the PDMS and the ϕ_{cmc} is ~ 0.002–0.3% in the PS. Muñoz et al. [33] described a two-step procedure for obtaining a PDMS/HDPE blend. First, a masterbatch containing a blend of 50 wt % of ultra high molecular weight PDMS with 50% LDPE is prepared. This masterbatch is then added up to 20 wt % to a HDPE matrix. Under these conditions, the polysiloxane is more easily dispersed in the polyolefin than when the PDMS is added alone to the HDPE. The resulting blend is then compatibilized using a grafted polyethylene. This is synthesized from HDPE, vinyltriethoxysilane (VTES) and dicumyl peroxide (DCP): Fig. 3. As shown by differential scanning calorimetry (DSC) measurements and SEM, the addition of 5 wt % of grafted copolymer into a 20 wt % masterbatch/75 wt % HDPE blend prevents the masterbatch domains from coalescing.

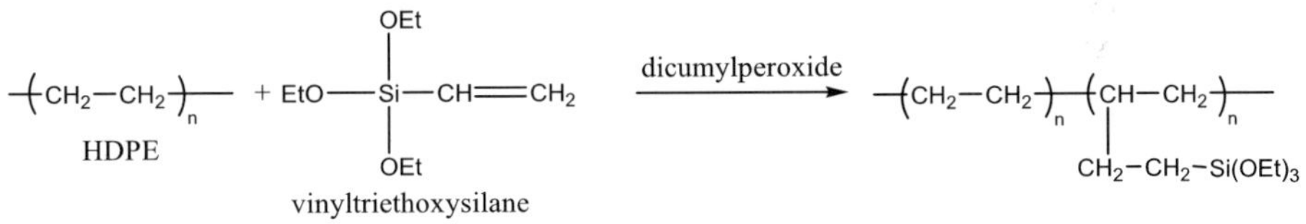

Fig. 3 Synthesis of HDPE-g-VTES compatibilizer

The chemical structures and architectures of the added copolymers have also been investigated. First of all, Wagner and Wolf [34] studied an A/B/A-B-A system consisting of PDMS and PEO homopolymers. They showed that the interfacial tension σ of a PEO/PDMS system decreases from 10.60 mN.m^{-1} without compatibilizer to about 0.70 mN.m^{-1} with a 2 wt % terpolymer (PDMS32-PEO37-PDMS32) concentration. For similar blends with the same composition, the PDMS block size increase induced a decrease in the value of σ (from 7.25 mN.m^{-1} for PDMS4-PEO37-PDMS4 to 0.70 mN.m^{-1} for PDMS32-PEO37-PDMS32), whereas the PEO block size had no effect on σ (e.g. 0.76 mN.m^{-1} for PDMS16-PEO14-PDMS16 and 1.16 mN.m^{-1} for PDMS16-PEO37-PDMS16). In both cases, a limiting value was observed beyond which the interfacial tension remained constant. Jorzik and Wolf also studied the effect of the copolymer architecture for the same system [35]. They compared diblock, triblock and "bottle-brush" copolymers (Fig. 4); the structure of the latter for example was poly(DMS21(MS(-EO)15))5-DMS). The order of the stabilization efficiency was found to be: triblock > "bottle-brush" > diblock copolymer. But, as pointed out by the authors, it should be remembered that experiments were limited to copolymer availability (molar mass, chemical composition, architecture), hence only trends and not absolute certainties were possible.

This is a surprising result since several theories and experiments based on other blends tended to show the opposite trend [36, 37], that is the diblock

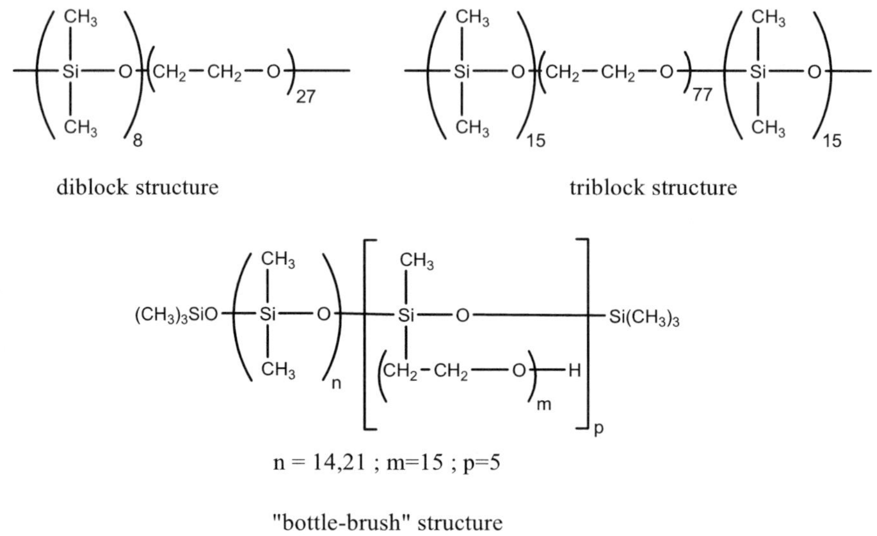

$$\text{–}\left(\underset{CH_3}{\overset{CH_3}{\underset{|}{\overset{|}{Si}}}}\text{–}O\right)_8\left(CH_2\text{–}CH_2\text{–}O\right)_{27}\text{–}$$

$$\text{–}\left(\underset{CH_3}{\overset{CH_3}{\underset{|}{\overset{|}{Si}}}}\text{–}O\right)_{15}\left(CH_2\text{–}CH_2\text{–}O\right)_{77}\left(\underset{CH_3}{\overset{CH_3}{\underset{|}{\overset{|}{Si}}}}\text{–}O\right)_{15}\text{–}$$

diblock structure   triblock structure

$$(CH_3)_3SiO\text{–}\left(\underset{CH_3}{\overset{CH_3}{\underset{|}{\overset{|}{Si}}}}\text{–}O\right)_n\left[\underset{(CH_2\text{–}CH_2\text{–}O)_m\text{–}H}{\overset{CH_3}{\underset{|}{\overset{|}{Si}}}}\text{–}O\right]_p\text{–}Si(CH_3)_3$$

n = 14,21 ; m=15 ; p=5

"bottle-brush" structure

Fig. 4 Jorzik et al. copolymer structures [35]

copolymers exhibited a better efficiency than the triblock copolymers. In the same paper [35], A/B/A-C and A/B/C-D systems were also studied. Two kinds of diblock copolymers were investigated: P(EO-b-S) and P(S-b-MMA), the polystyrene copolymer segments were located at the interface with the PDMS and PMMA copolymer segments at the PEO interface. They were found to be much less efficient than the PDMS-b-PEO copolymers.

3.2
In-Situ Copolymer Formation

In-situ copolymer formation has several advantages over the added copolymer approach. First, the preformed copolymers need to be dispersed during the blend mixing so that the dispersion efficiency ultimately depends on the blending technique. Next, there is the risk that the preformed copolymers micellize so that part of the added copolymer is no longer involved in the compatibilization process. On the other hand, since in-situ synthesized copolymers are already at the interface when they are formed, they are less inclined to form micelles and hence diffusion issues are avoided. An added advantage is the fact that in-situ-formed copolymers involve a one-step process. The main drawback of in-situ compatibilization is that it requires very reactive groups to be compatible with the residence time of common industrial processes. Although many reactive groups can be involved, anhydride/amine is up to now the best pair for in-situ copolymer formation since in addition to compatibilization [38, 39], it allows the preparation of nanostructured materials [40–46].

3.2.1
In-Situ Block Copolymer Formation

As will be discussed in the following section, although it is more usual to form block or graft copolymers by an in-situ process, this approach has not yet been extensively pursued in the case of polysiloxanes. Fleischer et al. [47] studied an A/B/A-B system which involved the formation of acid/base complexes (for example, carboxylic or sulfonic acids with tertiary amines) at the interface of the homopolymers. These complexes can be disrupted with a temperature increase. In the case of the following blends: PB, PDMS, PB-COOH and PDMS-NH_2, the authors specified that this mixture resulted in a "block-copolymer like" formation. Actually, the phenomenon that takes place is more a complexation that involves hydrogen bonding. This has been already studied in miscible systems such as poly(vinylphenol)/poly(vinyl acetate) that do not contain polysiloxane [48]. As in the case of the preformed copolymer addition, a linear interfacial tension reduction was observed for low concentrations of the added pair PDMS-NH_2/PB-COOH. For higher concentrations, a minimum was reached. In the case of the preformed copolymer, this plateau is observed beyond the cmc. The authors also showed that the interpolymer complex is dependent on the equilibrium between the free and associated end groups, that is to say, the stoichiometry. Hence, the complexation equilibrium reached at the beginning of the plateau was found to be dependant on the amino-acid equilibrium, and thus can be modified through the stoichiometry of the additives.

Maric et al. [49] also studied the in-situ block copolymer formation via reactive blending of functionalized homopolymers. In their work, blends were characterized by SEM, DSC and dynamic mechanical spectroscopy (DMS). It should be noted that their blends (PA-6/PDMS and PS/PDMS) were composed totally using functionalized homopolymers. The different reactions under investigation were: amine(NH_2)/anhydride(An), amine(NH_2)/epoxy(E) and carboxylic acid(COOH)/epoxy(E) (Fig. 5).

The morphologies of the different PA6/PDMS blends studied were found to be similar with a dispersed phase size of about 10 μm, 10.7 μm and 7.3 μm for unreactive PA6/unreactivePDMS, unreactive PA6/hydride functionalized PDMS and amine functionalized PA6/anhydride functionalized PDMS, respectively. The virtually unmodified rate of crystallinity and T_g values indicate that no significant reaction occurred during blending. SEC analyses of the PS/PDMS blends were performed since THF is a common solvent for these two polymers. PS/PDMS, amine-functionalized PS/epoxide-functionalized PDMS and CO_2H-functionalized PS/epoxide-functionalized PDMS were found to have similar coarse morphologies, and were unstable after annealing. But, the amine-functionalized PS/anhydride-functionalized PDMS morphology was very sharp (< 0.5 μm) and stable after annealing. The authors used SEC to determine the formation of 3% diblock copoly-

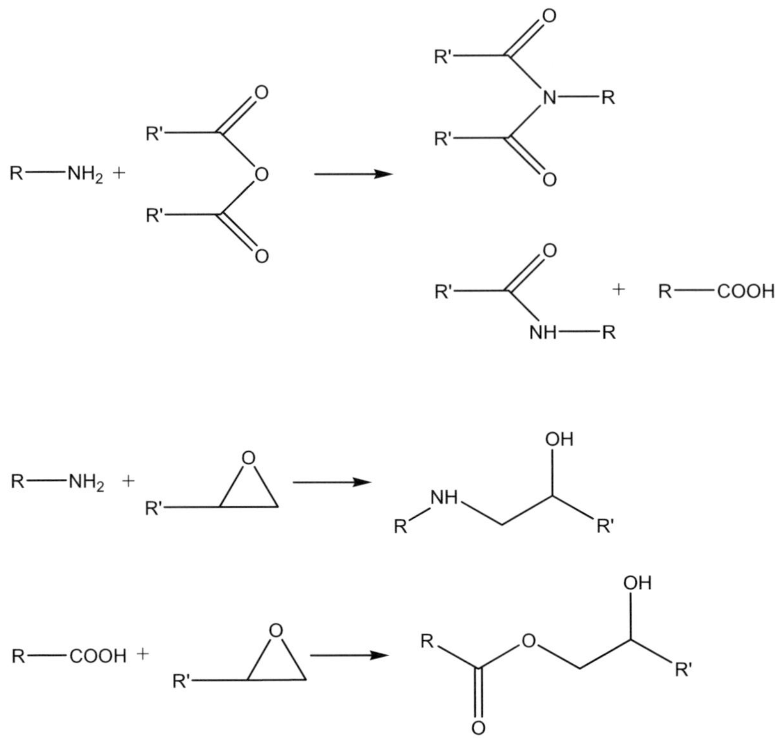

Fig. 5 Chemical reactions involved in work by Maric et al. [49]

mer volume fraction after mixing and noted that when these reactions were studied in other systems such as maleic anhydride containing poly(ethylene-co-propylene)/PA-6,6 [38], PS/PI or PS/PMMA [45], much more copolymer was obtained. They suggested that this is due to the high immiscibility of the two polymers so that the reaction can only take place within a very small area. It may be concluded that, although the concentration of resulting copolymer is less important when PDMS is involved in the blend, the amine/anhydride reactive pair gave the best results.

3.2.2
In-Situ Graft Copolymer Formation

In spite of the large number of grafted and side functionalized polysiloxanes commercially available and the variety of modification techniques available [25, 50] (hydrosilylation, thiol-ene chemistry, halogen substitution, polycondensation), only a few of them have been used as in situ-formed graft copolymer compatibilizers.

Several papers have appeared which deal with the use of poly(ethylene-*co*-methacrylate) which can be grafted onto a vinyl-containing PDMS via a first-order radical reaction [51–53] (Fig. 6).

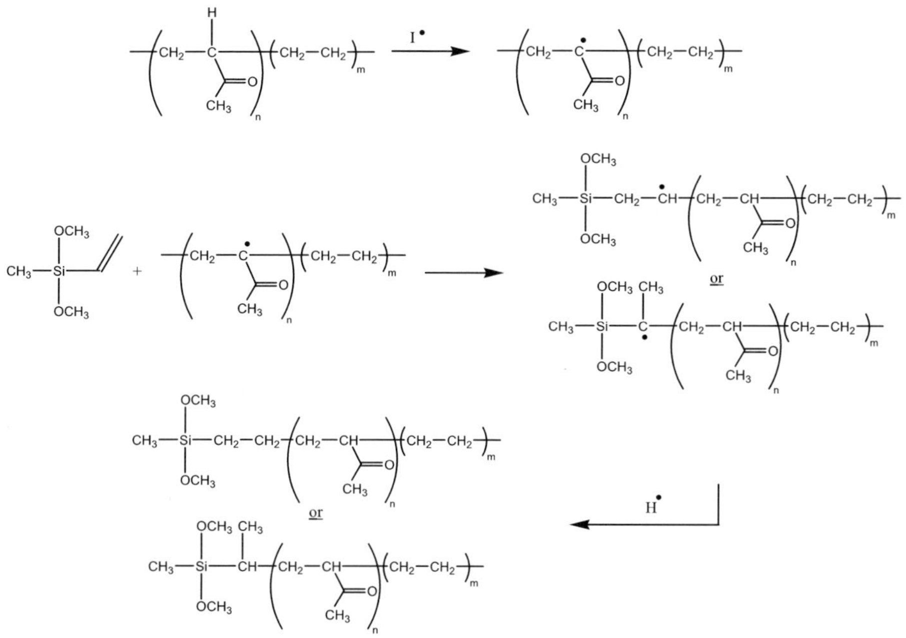

Fig. 6 Radical grafting of poly(ethylene-*co*-methacrylate) onto polysiloxane [52]

According to Nando et al., such graft copolymers allowed the in-situ compatibilization of polydimethoxysilane rubber and thermoplastics as well as low density polyethylene [54–56] and polyurethane [57]. Many different analytical methods (adhesion strength, SEM, impact strength, DMA, melt rheology, etc.) were used to determine that for a PDMS/LDPE (50 : 50) blend, a 6 wt % concentration of grafted poly(ethylene-*co*-methacrylate) was optimum. This composition was also found to give the best thermal stability to the blend [58]. For PDMS blends containing thermoplastic polyurethane, poly(ethylene-*co*-methacrylate) can interact with the latter via hydrogen bonding: the optimum concentration was found to be about 5 wt %.

Muñoz et al. [59] studied a PDMS-bearing a terminal methacryloxypropyl group. This end group functionality enabled the authors to graft on HDPE via a molten state radical process initiated by dicumylperoxide (DCP) in a batch mixer. The grafting reaction giving HDPE-*g*-PDMS was monitored via the evolution of the torque until a maximum was reached, meaning the end of the reaction was attained. The authors observed that compared to pure HDPE, there was a decrease in the crystallinity of the blends with the grafting rate increase.

3.2.3
In-Situ Branched Copolymer Formation

The two principal in-situ syntheses of branched copolymers are by step growth or radical chemistry. It should be noted that crosslinking of the same phase can also occur in addition to branching. This crosslinking is the basic principle of IPN formation. Hence, in this section, we will only refer to reports where crosslinking of the same phase appears to be a side reaction and not the expected one.

Shih et al. [60] studied the modification of a novolac-type epoxy resin with PDMS to overcome brittleness and poor impact resistance. This kind of resin is typically cured via their epoxy functions. The authors also introduced isocyanate monofunctionalized PDMS. Hence, the common treatment with MDA (4,4′-methylene dianiline) not only cured the resin on the one hand, but also made it possible to form the branched copolymer. Mechanical and thermal analyses showed that an optimum in isocyanate-terminated PDMS content was required to reach good thermal and physical properties and low moisture absorption.

Radical chemistry has been used for producing either PDMS/polyolefin blends or PDMS/unsaturated polymer blends. Grafting polyolefins via a radical mechanism is widespread due to the absence of usable functional groups and is particularly useful for silylated compounds [61–64]. Jana et al. [65] worked with LDPE and vinyl-containing PDMS which they cured with DCP (Fig. 7). Using DSC, they determined a first-order curing reaction. The proposed mechanism is depicted in Fig. 8. They also showed that the activation energy depends on both the amount of peroxide and the blend composition. Mechanical properties also varied with the DCP content and PDMS content. The optimum DCP content for the best tensile strength was found to be 1.5 wt %. Elongation at break decreased with increasing peroxide and PDMS.

Kole et al. [66, 67] studied a silicone-EPDM rubber blend. They enhanced the compatibility of the materials by the introduction of interactive [66] or reactive [67] functions grafted onto the two polymers. In both studies, PDMS

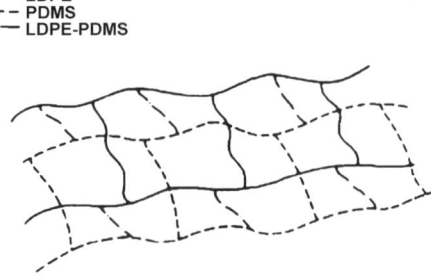

Fig. 7 Branched LDPE/ PDMS system, Jana et al. [65]

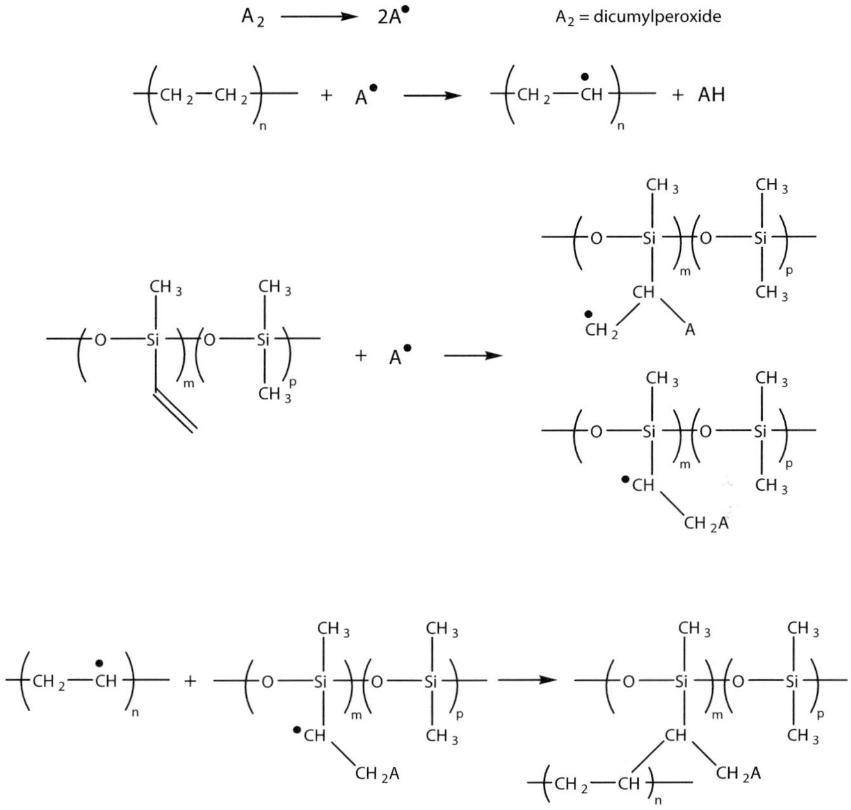

Fig. 8 Mechanism proposed by Jana et al. [65]

was grafted with polyacrylamide via a radical polymerization of acrylamide initiated from the PDMS backbone. In order to promote chemical interactions, EPDM was sulfonated, and for chemical reactivity, it was modified with maleic anhydride (MA-g-EPDM). On the one hand, for the interactive system PDMS-g-polyacrylamide/sulfonated EPDM, dipole-dipole and hydrogen bonding (Fig. 9) were identified by the FT-IR shifts of the carbonyl and amide stretching bands. It was observed that elongation at break was lower and the tensile strength was higher in this system than in the blend of the unmodified homopolymers. According to the authors, the improved ageing behaviour and thermal stability are due to the strong intermolecular forces. On the other hand, for the reactive PDMS-g-polyacrylamide /MA-g-EPDM system, a similar analysis showed the same tendencies that is to say a higher tensile strength and a lower elongation at break of the modified blend, and a much better retention of the properties vs. ageing and thermal stability. The use of a radical process to graft silicone has also been patented for fluoroelastomers/silicone blends [68, 69].

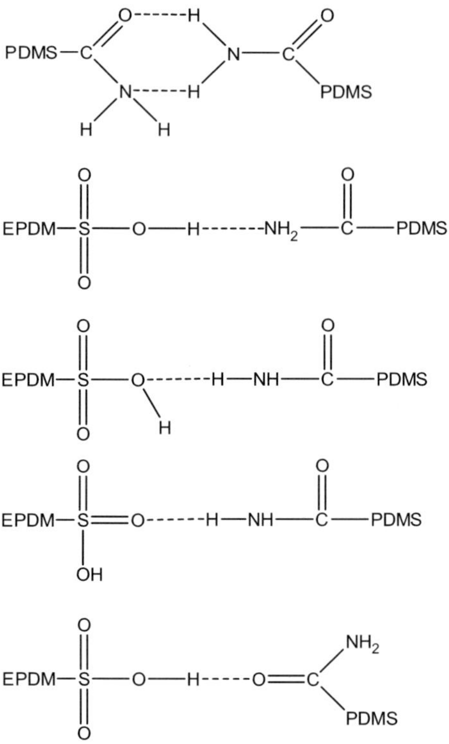

Fig. 9 Potential hydrogen bonding in PDMS-g-polyacrylamide/sulfonated-EPDM

An alternative way to generate radical species without adding initiator is by heating and shearing [70, 71]. The radical generation depends on the polymer structure, that is to say the relative ease of hydrogen abstraction from the backbone which increases with the degree of substitution of the carbons bearing labile hydrogens [72]. Jalali-Arani et al. [71] determined that LLDPE was more reactive than LDPE and that HDPE did not undergo radical grafting. In another report, Kole et al. [73] found that PP could be grafted with a vinyl methyl silicone rubber, but they observed that under similar conditions LDPE could not. As could be expected, when it occurred, grafting was found to increase with the temperature and the mechanical shearing rate that is to say with the level of generated radicals.

4
Interpenetrating Polymer Networks

Interpenetrating polymer networks (**IPNs**) are based on two (or more) polymers permanently entangled and independently crosslinked via specific reactions. The consequence of this imposed irremovable structure is the forced compatibilization of the polymers. This limits the phase separation via a segregation phenomenon that normally occurs when two incompatible polymers are blended. The phase separation limitation enables one to obtain materials with well-defined properties, for example mechanical properties which are required for many applications. Another example is the transparency which is directly dependant on the phase separation so that process has to be well controlled to obtain a phase domain size smaller than the light wavelength (about a few hundreds of nanometers) resulting in translucent materials. Although transparency is a useful indication of the quality of the interpenetration of the networks, Tg measurements (DMA, DSC, etc.) and more recently ^{13}C NMR cross-polarization give quantitative information. IPNs can be produced using different processes. On the one hand, simultaneous non-interfering polymerization and crosslinking leads to a **simultaneous IPN**. On the other hand, polymerization and crosslinking of the first polymer by forming the first network in the presence of the second monomer or prepolymer (by swelling or in solution) with subsequent polymerization and crosslinking in a second step leads to a **sequential IPN**. This can also be done in an emulsion media where the IPN obtained is referred to as a **latex IPN**. Overindulgently, the concept of interpenetration has been extended to blends where only one of the two components form a network, the other one being a non-branched polymer. These blends are called **semi-IPNs** (occasionally named **pseudo-IPNs**).

4.1
Silicone Crosslinking Methods

Silicone crosslinking can be obtained via different reactions which have been classified by Brook into six categories [74]: (1) "alcoholysis", (2) room tem-

Table 1 Polysiloxane crosslinking systems

	Cure system	Chemical entities	Catalysts
1	"alcoholysis"	$R_3SiH + HOSiR'_3$	Pt complex, tin carboxylates
2	RTV: room-temperature vulcanization	Silanol + $RSiX_3$ or SiX_4 X = Cl, carboxylates, oximes, alkoxides	Tin, titanate, acidic or basic catalysts
3	HTV: high temperature vulcanization	Polysiloxane or vinyl containing polysiloxane	Radical generator (chiefly peroxide)
4	Radiation and UV	Polysiloxane or vinyl containing polysiloxane	
5	Hydrosilylation [117]	R-SiH + vinyl group	Pt based complexes[a]

[a] Other metal complexes (rhodium, palladium ...) are seldom used in our IPN synthesis

perature vulcanization (RTV), (3) high-temperature vulcanization (HTV), (4) radiation and UV-induced crosslinking, (5) hydrosilylation cure and (6) crosslinking via organic side chains (Table 1).

Many silicones grafted with organic entities are commercially available and others can be synthesized [50] giving rise to numerous organic crosslinking methods, only those involved in IPNs will be discussed in this review.

4.2
IPNs Synthesis

PDMS use in the IPN field is very attractive since it allows production of materials with good mechanical properties. Although, as will be described, depending on the targeted applications, some drawbacks of the silicones have to be overcome, such as either their hydrophobicity or their permeability. The silicone-containing IPNs are presented by preparation methods to highlight the PDMS chemistry and any issues of the processes. Many reports are dedicated to the point of view of application thanks to the useful properties of PDMS, however, the ins and outs such as kinetics, morphology or structure/property relationships are also pointed in this discussion.

It should be kept in mind that networks can be obtained from either monomers or from low molecular weight prepolymers. In the following discussion, "monomer" will be understood to cover either monomer or prepolymer.

4.2.1
Sequential IPNs

There are two principal approaches to forming sequential IPNs: (1) form the first network, swell it with the second monomer, crosslinker and catalyst and then form the second network; (2) blend the two monomers, crosslinkers and catalysts together and then crosslink them. Two different initiation processes (e.g. different temperatures) can be used in what is called **in-situ sequential synthesis**. Finally, an alternative consists in blending the monomers and then adding the catalysts and/or the crosslinkers sequentially.

The advantage of the "swelling method" is that it is not limited by the crosslinking reactions of each phase so any interference from these will be limited. A good representative example is the synthesis developed by Hamurcu and Baysal [75]. They synthesized a bimodal PDMS (15 000 g mol^{-1})/PDMS (75 000 g mol^{-1}) IPN with the same condensation curing system. First, the 75 000 g mol^{-1} PDMS network was formed from the corresponding α,ω-dihydroxypolydimethylsiloxane and tetraethylorthosilicate catalyzed by stannous 2-ethylhexanoate. It was then swollen in a 15 000 g mol^{-1} α,ω-dihydroxy-poly(dimethylsiloxane) monomer. The second monomer was then crosslinked via the same condensation cure. The sequential full IPN structure was confirmed by DSC analysis which exhibited two distinct Tgs (− 128.1 °C and − 122.3 °C).

Many polyacrylate/PDMS IPNs are synthesized via such a swelling method. Some of these systems are silicone rubber/hydrogels. They are extensively used when the application requires hydrophilicity of the rubber. With this end, the most reported monomers are methacrylic acid and 2-hydroxyethylmethacrylate. Turner et al. [76, 77] developed a typical synthesis to form a PDMS/polymethacrylic acid (PMAA) IPN via a two-step process. First, a PDMS film was prepared by hydrosilylation of vinyl-terminated PDMS by tetramethylcyclotetrasiloxane (a hydride crosslinker). Then, the mixture of components plus a platinum catalyst were sprayed onto an appropriate support and afterwards cured at 55 °C to obtain a "pre-IPN film". The authors studied the influence of the second step (MAA polymerization) on the material morphologies and properties. Co-continuous structures are known to be obtained at high concentrations of the host monomer and the so-called "sea-island morphologies" (nucleation and growth phase separation mechanism) are obtained for low concentrations of the host monomer. Co-continuous structures lead to permeable materials whereas "sea-island morphologies" lead to impermeable materials. Depending on the targeted applications, one of these two morphologies is needed. The authors [77] prepared different rubber-hydrogel IPN membranes and tested their permeation to water soluble components (vitamin B12). They concluded that when the crosslinked PDMS film is swollen by MAA monomer and crosslinker (triethylene glycol dimethacrylate and UV sensitive free radical initiator) and disposed onto a support to be UV-cured, a gradient of concentration of the monomer in the PDMS film occurred. This gradient can be attributed on the one hand to the nature of the interface with the support (glass in this study which has specific interactions with the MAA) and on the other hand with the air. Moreover, some evaporation of the monomer undoubtedly occurred during handling thus increasing the gradient. The IPNs in these cases were impermeable to water soluble components such as vitamin B12 since the low concentration of monomer on one side of the film leads to a "sea-island morphology" forming locally. These low MAA concentrations were observed as expected either opposite the glass-IPN interface or at the air–IPN interface. The authors developed an alternative method: the crosslinked PDMS film was submerged into the mixture of monomer, initiator and crosslinker and once it was swollen, it was UV-cured. In this case, no gradient was observed and hence the membrane was permeable to vitamin B12. The influence of initiator and crosslinking agent concentrations, UV light intensity and the IPN morphology on the permeation properties were also studied [78, 79]. A similar material was patented [80] for contact lens applications since the final material is sufficiently hydrophilic thanks to the PMAA and is also oxygen permeable. Poly(2-hydroxyethyl methacrylate) was also combined with PDMS to obtain a hydrogel/silicone IPN (poly(HEMA)/PDMS) [81]. This system has been extensively studied because of its biocompatibility and is widely used in the medical field. But it was mainly prepared by blending a crosslinked poly(HEMA) powder with vinyl-containing silicone and radical initiator. Afterwards polysiloxane crosslinking was achieved by heating the blend [82–89]. Although these materials, which are not IPNs, are called silicone/hydrogel composites, the latter examples have been compared to real poly(HEMA)/PDMS IPN equivalent [90]. Tg measurements showed that phase separation was a little bit less important for the real IPNs. In the case of the IPN structure, tensile strength and elongation at break were enhanced whereas the hardness diminished. The preparation of a poly(HEMA) hydrogel/PDMS rubber IPN via mixing all the components and curing the mixture in two temperature steps was also claimed [91].

IPNs are also attractive for development of materials with enhanced mechanical properties. As PDMS acts as an elastomer, it is of interest to have a thermoplastic second network such as PMMA or polystyrene. Crosslinked PDMS have poor mechanical properties and need to be reinforced with silica. In the IPNs field, they can advantageously be replaced by a second thermoplastic network. On the other hand, if the thermoplastic network is the major component, the PDMS network can confer a partially elastomeric character to the resulting material. Huang et al. [92] studied some sequential IPNs of PDMS and polymethacrylate and varied the ester functionalities: the polysiloxane network was swollen with MMA (methyl methacrylate), EMA (ethyl methacrylate) or BuMA (butyl methacrylate). Using DMA the authors determined that the more sterically hindered the substituent, the broader the damping zone of the IPN (Table 2). This damping zone broadness was also found to be dependant on the PDMS content, and atomic force microscopy (AFM) was used to observe the co-continuity of the IPN.

Silicone/PMMA IPNs have been prepared by crosslinking the silicone network at room temperature (RTV) and crosslinking the PMMA at a higher temperature. He et al. [93–95] crosslinked the silicone via the condensation of a disilanol and tetraorthosilicate catalyzed by stannous octoate. The second network was formed by radical polymerization of methyl methacrylate with

Table 2 Damping zone observed for different PDMS/Polyacrylate IPNs [92]

Composition (wt. %)	Damping temperature range of T_g (°C)		
	T_{onset}	T_{end}	ΔT
PDMS/PAC/PMMA 37.0/37.0/26.0	– 50	170	220
PDMS/PAC/PEMA 36.2/36.2/27.5	– 50	120	170
PDMS/PAC/PBuMA 34.4/34.4/31.2	– 35	120	155

trimethacrylate (present in the mixture before silicone crosslinking) at the higher temperature. They interestingly compared "full IPNs" to "graft IPNs". The difference between these materials is in the covalent bond between the PDMS and PMMA networks in the second case thanks to a PDMS crosslinker bearing a methacrylic function. The authors observed that Young's modulus increased from 2.1 for pure PDMS to 47 MPa for a PDMS/PMMA graft IPN (70/30 wt %) and decreased from 3.0 to 1.4 for the corresponding full IPN. Also, elongation at break increased from 26 to 200% for full IPN, whereas it remained unchanged for graft IPN. The polymethylphenylsiloxane (PMPS)/PMMA system was also studied and patented [96–98]. Caille et al. [96] studied a silicone PMPS/PMMA IPN. The advantage of PMPS is that its refractive index is equal to that of the PMMA, and hence it preserves the transparency property of the final material. Irrespective of the preparation of the IPNs (either crosslinking the PMPS or PMMA first), the material obtained did not exhibit any IPN properties. Subsequently a "mixed" crosslinker—trimethoxysilylpropylmethacrylate—was copolymerized with PMMA and PMPS, resulting in a material which revealed IPN-like properties. In the same way, Brachais et al. [97] studied a PMPS/PMMA IPN where analysis showed that the PMPS tended to belong simultaneously to the rich PMPS phase as well as to the PMPS/PMMA interpenetrating domains. The PMMA on the other hand was more homogeneously dispersed in the PMPS/PMMA interpenetrating domains. The authors suggested that during the second network formation at 70 °C, the PMPS chains which were not yet entangled formed these PMPS-rich domains. This phenomenon is facilitated because of the large difference between the process temperature and the low Tg of the polysiloxane making some chain movements possible. Miyata et al. [99] used the same process to prepare a PDMS/PS IPN where the PS network was formed by copolymerization of styrene and divinylstyrene at 80 °C. The authors choose these two polymers because of their permselectivity. In fact, PS is used in membranes for its ethanol-permselectivity whereas PDMS has more affinity for ethanol than for water. Hence, permeation of aqueous ethanol solutions through this IPN membrane was as efficient as through PS membranes but was mechanically much better. Tsumara and Iwahara [100] prepared LDS/PCS IPNs—LDS being a ladder silsesquioxane and PCS a polycarbosilane (Fig. 10). Hence, the first LDS network was crosslinked at room

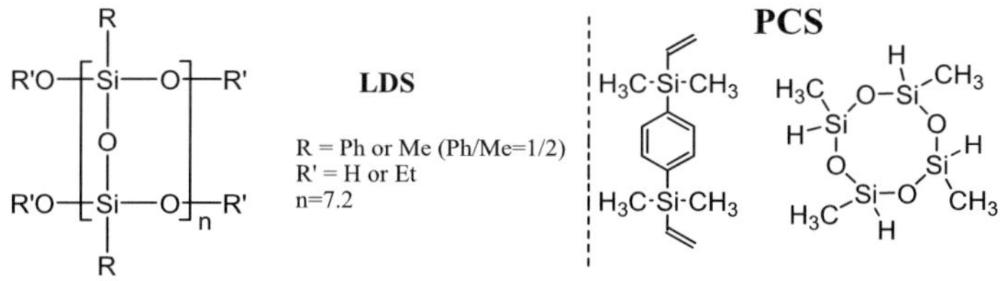

Fig. 10 Tsumara and Iwahara network starting reagents [100]

temperature (RTV) whereas the second one was formed at high temperature. Synergistic properties were found since the mechanical properties were better than those of either LDS or PCS alone. Moreover, the IPN exhibited good optical and electrical properties.

Fichet et al. [101] studied a PDMS/cellulose acetate butyrate (CAB) IPN. CAB is an OH-functionalized environmentally friendly polymer, transparent and relatively cheap. It was crosslinked using Desmodur® N3300 (a NCO functionalized compound) and dibutyltin dilaurate. It is interesting to notice that the vinyl-containing PDMS was crosslinked via a thiol-ene system—the crosslinker was a trithiol: trimethylolpropane tris (3-mercaptopropionate)—initiated by AIBN. DMA of the IPNs revealed a single T_g and no phase separation confirming a well-interpenetrated material. Darras et al. [102] synthesized a fluorinated polysiloxane-based IPN. A PDMS network was first obtained by treatment of disilanol with Desmodur® N3300 at 40 °C and then the fluorinated network was formed by treatment of a polyperfluoroalkyl acrylate with ethylene glycol dimethacrylate (EGDMA) and AIBN at 55 °C. The IPN was transparent and its mechanical properties were enhanced. It should be noted that the simultaneous formation of the two networks or the formation of the fluorinated networks before the silicone one gave phase separated blends.

As can be seen from all of these works, sequential IPNs have been extensively used with silicone with various crosslinking reactions and polymers. As a result, attention must be paid to the influence of processing conditions, of the chemical nature of components (silicone and the other polymer), of the evolution of the material at each step of the IPN synthesis etc. Depending on the mastering of these parameters, reactivity, morphology and entanglement can be controlled and consequently targeted properties can be reached.

4.2.2
Simultaneous IPNs

The main difficulty in obtaining a simultaneous IPN is to find two non-interfering crosslinking reactions which allow the formation of the desired polymer networks. Generally, the two monomers are vigorously mixed and, once a good mixing is achieved, then the crosslinkers and initiators are added and crosslinking takes place. Three main factors drive the simultaneous IPN morphology. Before the crosslinking occurs, the mixing of the monomers has to be as efficient as possible and this is directly linked to the compatibility of the first polymer with the second (this is often expressed by the solubility parameter difference $\Delta\delta$). Once the polymerization and crosslinking have begun, the difference between the T_g's of the two polymers is important because it drives the difference in the mobility of the chains. If one polymer has a high chain mobility, the crosslinking will occur too late and entanglement will not be good (the same phenomenon explains the rich PMPS phase in the sequential IPN of Brachais [97] described above). This difference in T_g's is extremely important in the case of silicone since this polymer has a very low T_g. Finally, the IPN morphology also depends on the kinetics of the formation of the networks: the more simultaneous the crosslinking, the more entangled the IPN. As described earlier, silicone crosslinking can essentially be made through hydrosilylation or condensation. Concerning the case of the PPO/PDMS IPN (Frisch and Gebreyes [103]), crosslinked hybrid-terminated silicone with tetraallyloxysilane (hydrosilylation initiated by AIBN and brominated PPO was crosslinked using ethylenediamine. Both reactions occurred at 40–70 °C. The mechanical properties were studied with respect to variation of the PDMS content: tensile strength decreased and elongation at break increased when the PDMS ratio increased. A similar PPO/PDMS system was prepared by forming the silicone network with hydroxy-terminated polydimethylsiloxane and tetraallyloxisilane condensation catalyzed by stannous ethyl hexanoate [104]. The IPNs obtained exhibited a single T_g for a composition of up to 40 wt % of silicone which could be associated with a single phase morphology. TEM analysis tended to support this hypothesis [105]. So Ebdon, Hourston and Klein [106] crosslinked a silicone [α, ω-hydroxy-terminated poly(dimethylsiloxane)] with tetraethoxysilane (TEOS) and dibutyltin dilaurate while an isocyanate-terminated polyetherurethane was crosslinked with trimethylolpropane. Unfortunately, these IPNs did not exhibit any significant mechanical property improvements. The same conclusion was reached for the PDMS/ESE IPN where ESE represents a hydroxy-terminated POE-PDMS-POE triblock copolymer crosslinked with TDI while PDMS was crosslinked thanks to its end hydroxyl groups catalyzed by TEOS and dibutyltin dilaurate [107]. The use of PMPS rather than PDMS enabled Klein et al. [108] to decrease the phase separation between the silicone and the other polymer. This was as expected when the solubility parameters of the polymers are considered. In each case, they observed a continuous phase of polyetherurethane with a dispersed phase of a certain amount of silicone depending on the polymer structure. In poly(dimethylsiloxane-urethane)(PDMSU)/PMMA IPNs, Zhou et al. [109, 110], observed that the degree of crosslinking drives the maximum amount of PDMSU for which the phase is continuous.

4.2.3
Latex IPNs and Latex Semi-IPNs

A few papers dealing with silicone latex IPNs have appeared. Frisch et al. [111] patented a process whereby two emulsions were prepared: the first contained a hydroxyl-terminated PDMS and stannous octoate, the second a crosslinked polyurethane, poly(urethaneurea) or polyacrylate latex. The two lattices were blended and a film was cast and cured at 120 °C. The mechanical properties were found to be enhanced, especially in the case of the polyacrylate/PDMS IPN. Noteworthy is the fact that semi-IPNs such as the thermally stable

PTFE/PDMS can be obtained via a similar process [112]. PTFE chains are trapped in the PDMS network during casting. Jones et al. [113] also patented a process where adipic acid and hexane diamine (or ε-caprolactam [114]) were added to a silicone latex leading to the formation of a polyamide/PDMS semi-IPN.

4.2.4
Semi-IPNs

From the considerable number of patents in this category, it appears that semi-IPNs are of particular commercial interest. Thermoplastic elastomers (TPE) where the elastomer phase is crosslinked are sometimes called thermoplastic vulcanizates (TPV). More specifically, when the elastomer phase is a silicone one, then the material is called a silicone thermoplastic vulcanizate (**STPV**) [115]. One can also find the term "TPSiV" used but it is a trade name and hence is less employed. This growing commercial interest is especially due to the characteristics of the final material which simultaneously exhibit the properties of both the thermoplastic and the silicone. Moreover, their processability and recycling potential make them of great interest compared to strict IPNs. As the network formation forces the polymers to entangle, chemical interactions should play a minor role. It has nevertheless been observed that the chemical nature of the silicone side groups drives the degree of interpenetration [116] and hence the phase separation and the properties of the final materials. Among the other parameters that drive the final structure are the crosslink density and method of synthesis, both of which are of prime importance.

The original use of hydrosilylation [117] in semi-IPN formation is attributed to Arkles [118–120]. His approach was to mix two polysiloxanes with the thermoplastic. One contained hydride functions and the other vinyl functions. The subsequent introduction of a platinum catalyst induced crosslinking. Depending on the number of hybrid and vinyl functions per chain, either chain-extension or crosslinking was observed. As pointed out by Arkles [119], according to theory, these phenomena should be obtained for a 1:1 hydride:vinyl group ratio, however, in practice the hydride functions were consumed by several side reactions so that a higher hydride ratio was required.

This technology was first commercially applied to polyurethane blend [121] and patented as Rimplast® (for Reactive Injection Molding), but many polymers have since been blended with polysiloxane thanks to this method: polyethylene [122], polypropylene [122, 123], polyamide [124–130], polyesters [128, 131–133], poly(phenylene ether) [134], fluorocarbons [135] and many more. Many of them include reinforcing fillers such as fumed silica. The silicone base involved can moreover contain reactive groups such as the epoxy group [136, 137]. A typical silicone base useful for these blends was described in Bilgrien's patent [138] resulting in the mechanical properties and processability of the semi-IPN being enhanced. Besides, the incorporation of this silicone base conferred on the semi-IPN an improvement in the burn character [139].

It is well recognized that side reactions can occur during the process between the different compounds present in the media due to drastic environment particularly high temperatures and mechanical shearing. For example, Fustin et al. [140] studied a melt-reacted polyester/silicone system (200 °C in sealed vials). The silicone was obtained by hydrosilylation catalyzed by Karstedt's platinum catalyst. To make the analysis easier, PBT was mimicked by low molecular weight compounds which contained potentially reactive groups towards the hydride group of the silicone (ester, acid, alcohol, alkene or α, β-unsaturated ester). The behaviour of each group towards the hydrosilane was individually studied (Fig. 11). When the hydrosilane was heated alone, it partially reacted by itself to lead to a dimer. No reaction was observed between the hydride-containing silicone and the ester groups. With acid (to a minor extent) and alcohol groups, cross dehydrocoupling was observed. The hydrosilylation reaction of the α, β-unsaturated ester and alkene were found to be faster than cross-dehydrocoupling with the alcohol. It was found that hydrosilylation between the hydride-functionalized silicon and the vinyl-containing silicon was faster than any other potential side reaction. Moreover, dimerization of the hydrosilane was found to be slower than the hydrosilylation reactions.

$$R_3SiH \xrightarrow{\text{dimerization by hydrocoupling}} R_3Si\text{-}SiR_3 + H_2$$

$$R'COOH + R_3SiH \xrightarrow{\text{cross-dehydrocoupling}} R'COO\text{-}SiR_3 + H_2$$

$$R'OH + R_3SiH \xrightarrow{\text{cross-dehydrocoupling}} R'O\text{-}SiR_3 + H_2$$

$$R_3SiH + R'Si\text{-}CH=CH_2 \xrightarrow{\text{hydrosilylation}} R_3Si\text{-}CH_2\text{-}CH_2\text{-}R'$$

Fig. 11 Fustin et al. reactions [140]. Catalyst: Pt(0)

It should be noted that the dimerization side reaction is hugely diminished when a radical scavenger such as Irganox 1010 is added to the medium. The latter one of a group of convenient hindered phenol stabilizers, but thioesters and hindered amine have also been incorporated to the blends [131].

Itoh et al. [141] patented a blend of polyurethane and silicone gum with vinyl groups. The crosslinking reaction of the silicone component was achieved using a radical generator. Chorvath et al. [142] also patented a system composed of polyamide or polyester as the thermoplastic and vinyl-containing polysiloxane cured by a peroxide or non-peroxidic initiator such as 3,4-dimethyl-3,4-diphenylhexane (HTV). Moreover, they added different compatibilizers such as amino-terminated or amino-grafted polysiloxane, epoxy- or isocyanato-functionalized polysiloxane.

Another way to achieve crosslinking of the silicone is via condensation. Fu and Mark [143] obtained a silicone network by mixing hydroxy-terminated PDMS and tetraethoxysilane. The reaction was catalyzed by a stannous compound. The network obtained was then swollen with styrene whose polymerization was initiated by benzoyl peroxide. Ultimate strength and elongation at break were both enhanced. This crosslinking system has also been used in other patented systems involving for example poly(arylene sulphide) [144, 145]. Fujiki et al. [146] prepared silicone/PMMA semi-IPNs where the silicone is crosslinked by condensation but with an acidic catalyst to avoid the use of organotin compounds. For example, a semi-IPN was obtained by mixing together hydroxyl-terminated-PDMS, tetraethoxysilane, an acidic catalyst such as acetic acid, MMA monomer and AIBN before heating and curing.

Arkles [147] combined the hydrosilylation and condensation mechanism to obtain a secondary crosslinked semi-IPN. The polyamide thermoplastic was blended with a vinyl- and alkoxy-containing silicone. The hydrosilylation first occurred to yield a semi-IPN. In a second step, this semi-IPN was exposed to moisture and underwent the second crosslinking stage. Gilmer et al. [148] prepared a silicone network/PMMA system by condensation of chlorosilane and/or silanol. The mechanical properties were studied and the IPNs were found to be tougher when the degree of crosslinking was higher. The silicone was based on an aliphatic and aromatic silane, so that the authors obtained an IPN with an aromatic silane which had a transmittance of 79% whereas that of PMMA is 92% (aliphatic silicone-based IPN has 0% transmittance).

Finally, the "alcoholysis" ($RSiOH + HSiR'$) catalyzed by metal carboxylate, amine or quaternary ammonium was patented by Gornowicz [149]. Various thermoplastics such as PA, PP, SBS and PE were described in the different examples.

4.2.5
Mixing Method

To enhance the efficiency and the stability of the semi-IPNs, their preparation can include a second system of compatibilization.

4.2.5.1
Preformed Compatibilizer

Some authors combined the IPN concept with the use of compatibilizers similar to that mentioned in Section 2. These may be a coupling agent such as epoxy-functionalized polysiloxane, polysiloxane copolymers or an organofunctional grafted polyolefin such as poly(ethylene-co-methacrylate) or MA-g-EPDM (maleic anhydride grafted EPDM) [124, 125, 133]. Knaub et al. [150] studied a poly(urethane-ureas)/PDMS semi-IPN in which a poly(urethane-*co*-dimethylsiloxane) block copolymer was added as a compatibilizer for the two phases. A battery of tests (DMA, SEM, stress-strain measurements) showed an enhancement of the semi-IPN properties for 0.5 to 2 wt % amounts of added copolymer. For higher concentrations, the mechanical properties diminished compared to those of the unmodified network. This is due to the fact that the additive acts like a plasticizer. Vlad et al. [151] used a polyalkylene oxide grafted PDMS to compatibilize a PU/PDMS IPN but its influence was not clearly pointed out. Gross et al. [133] added an ethylene-methylacrylate-glycidyl methacrylate terpolymer to a PBT/PDMS blend which may act as a compatibilizer.

4.2.5.2
Grafted Compatibilizer

Gornowicz [135] grafted a fluorocarbon resin (e.g. either a homo- or copolymer of VDF) with siloxane to enhance its compatibility with PDMS. For example, vinyl triethoxysilane was first grafted onto fluorocarbon resin using peroxide. Then, an added silicone phase was vulcanized via hydrosilylation.

4.2.5.3
Branched Semi-IPNs

Xiao et al. [152] prepared a PU/PDMS semi-IPN by crosslinking into the PU component a hydroxy-terminated PDMS bearing an amine function such as 3-aminopropyltriethoxysilane with a trialkoxysilane and dibutyltindilaurate. Unfortunately, the system appeared to be phase separated. Similarly, Morin [153] studied a polyamide/PDMS semi-IPN. A vinyl-containing polysiloxane was mixed with a polyamide. Then, a hybrid-containing silicone bearing epoxy functions was added to the blend to crosslink the silicone phase. The epoxy groups act as a reactive coupling agent with the PA phase. A semi-IPN of this grafted PDMS with Nylon 6,6 exhibited enhanced properties.

4.2.5.4
Hybrid Semi-IPNs

Hydrosilylation can also be carried out between a silicone elastomer and a non-silicone one to yield a hybrid IPN. Arkles [154] worked with such a system: the silicone containing hydride groups and the unsaturations supplied by polymers such as polybutadiene or copolymers of butadiene. These hybrid systems do not perform as well as siloxane semi-IPNs due to the fact that silicone is partially lost.

Finally, it should be pointed out that hydrosilylation crosslinking during melt mixing is so efficient that many thermoplastic elastomers where a section of the elastomer is a non-silicone one—such as EPDM —can be obtained via a hybrid-containing silicone crosslinking agent [155–157].

5
Conclusion

Despite the incompatibility of silicone towards many polymers, several techniques have been more or less successfully developed to produce silicone containing physical blends. Although compatibilization remains essentially of academic interest only, many industrial applications for IPNs are a testimony to their growing importance. Silicone blends allow researchers to introduce specific properties such as impact resistance or low surface energy to polymers that fundamentally lack these characteristics. On the other hand, it must be acknowledged that silicone blends and related fields of research are still open to further development and will undoubtedly lead to a wide range of future industrial applications.

Acknowledgements We gratefully acknowledge Dr. Edeline Wentrup-Byrne (Queensland University of Technologie, Brisbane, Australia) for her kind help.

References

1. Goldberg EP (1961) Resinous mixtures of polysiloxanes and polymers from carbonates of dihydric phenols. US Patent 2 999 835
2. Bostick EE, Jaquiss DBG (1973) Compatible polycarbonate-siloxane composition. US Patent 3 751 519, General Electric, US, p 2
3. Meyer RV, Dhein R, Fahnler F (1979) Polyamide blends with high impact strength, DE Patent 2 734 693
4. Hill DJT et al. (1996) Development of wear-resistant thermoplastic polyurethane by blending with poly(dimethyl siloxane). I. Physical properties. J Appl Poly Sci 61(10):1757–1766
5. Bremner T et al. (1997) Development of wear-resistant thermoplastic polyurethanes by blending with poly(dimethyl siloxane). II. A packing model. J Appl Poly Sci 65(5):939–950
6. Damrongsakkul S, Sinweeruthai R, Higgins JS (2003) Processability and chemical resistance of the polymer blend of thermoplastic polyurethane and polydimethylsiloxane. Macromolecular Symposia. 7th Eur Symp Polymer Blends, Lyon-Villeurbanne, 27–29 May 2002, pp 411–419
7. Wu S (1987) Formation of dispersed phase in incompatible polymer blends: interfacial and rheological effects. Poly Eng Sci 27(5):335–343
8. Maric M, Macosko CW (2002) Block copolymer compatibilizers for polystyrene/poly(dimethylsiloxane) blends. J Polym Sci Part B: Polym Phys 40(4):346–357
9. Chuai CZ et al. (2004) The effect of compatibilization and rheological properties of polystyrene and poly(dimethylsiloxane) on phase structure of polystyrene/poly(dimethylsiloxane) blends. J Polym Sci Part B: Polym Phys 42(5):898–913

10. Pötschke P, Paul DR (2003) Formation of co-continuous structures in melt-mixed immiscible polymer blends. J Macromol Sci Part C: Poly Rev C43(1):87–141
11. Avgeropoulos GN et al. (1976) Heterogeneous blends of polymers. Rheology and morphology. Rubber Chem Technol 49:94
12. Miles IS, Zurek A (1988) Preparation, structure, and properties of two-phase co-continuous polymer blends. Poly Eng Sci 28:796
13. Jordhamo GM, Manson JA, Sperling LH (1986) Phase continuity and inversion in polymer blends and simultaneous interpenetrating networks. Poly Eng Sci 26:517
14. Paul DR, Barlow JW (1980) J Macromol Sci Part C: Poly Rev C18:109
15. Utracki LA (1991) On the viscosity-concentration dependence of immiscible polymer blends. J Rheol 35(8):1615–1637
16. Anastasiadis SH, Gancarz I, Koberstein JT (1988) Interfacial tension of immiscible polymer blends: temperature and molecular weight dependence. Macromolecules 21(10):2980–2987
17. LeGrand DG, Gaines GL Jr (1969) Molecular weight dependence of polymer surface tension. J Colloid Interf Sci 31(2):162–167
18. LeGrand DG, Gaines GL Jr (1973) Surface tension of homologous series of liquids. J Colloid Interf Sci 42(1):181–184
19. Lee MH et al. (2001) The effect of end groups on thermodynamics of immiscible polymer blends. 2. Cloud point curves. Polymer 42(21):9163–9172
20. Fleischer CA et al. (1993) The effect of end groups on thermodynamics of immiscible polymer blends. 1. Interfacial tension. Macromolecules 26(16):4172–4178
21. Patterson HT, Hu KH, Grindstaff TH (1971) Measurement of interfacial and surface tensions in polymer systems. J Poly Sci, Polymer Symposia 34:31–43
22. Fleischer CA, Koberstein JT (1990) The effect of polymer end groups on the compatibility of immiscible polymer blends. Polymer Preprints (American Chemical Society, Division of Polymer Chemistry) 31(2):541–2
23. Furukawa H, Shirahata A (1994) Polyamide resin composition, EP Patent 581 224
24. Li XG, S H, Lai YH, Wee ATS (2000) Miscibility of carboxyl-containing polysiloxane/poly(vinylpyridine) blends. Polymer 41:6563–6571
25. Belorgey G, Sauvet G (2000) Organosiloxane block and graft copolymers. In: Silicon-Containing Polymers. Kluwer, Rotterdam, pp 43–78
26. Yilgor I, McGrath JE (1988) Polysiloxane-containing copolymers: a survey of recent developments. In: Advances in Polymer Science 86 (Polysiloxane Copolm/Anionic Polym). Springer, Berlin Heidelberg New York, pp 1–86
27. Macosko CW et al. (1996) Compatibilizers for melt blending: Premade block copolymers. Macromolecules 29(17):5590–5598
28. Cho D et al. (2000) Segregation dynamics of block copolymers to immiscible polymer blend interfaces. Macromolecules 33(14):5245–5251
29. Biresaw G, Carriere CJ, Sammler RL (2003) Effect of temperature and molecular weight on the interfacial tension of PS/PDMS blends. Rheologica Acta 42(1–2):142–147
30. Leibler L (1988) Emulsifying effects of block copolymers in incompatible polymer blends. Makromol Chem, Marcomol Symp 16:1–17
31. Hu W et al. (1995) Interfacial tension reduction in polystyrene/poly(dimethylsiloxane) blends by the addition of poly(styrene-b-dimethylsiloxane). Macromolecules 28(15):5209–5214
32. Chuai C et al. (2004) Influence of diblock copolymer on the morphology and properties of polystyrene/poly(dimethylsiloxane) blends. J Appl Poly Sci 92(5):2747–2757

33. Munoz PMP et al. (2002) Blends of high-density polyethylene with solid silicone additive. J Appl Poly Sci 83(11):2347–2354
34. Wagner M, Wolf BA (1993) Effect of block copolymers on the interfacial tension between two immiscible homopolymers. Polymer 34:1460–1464
35. Jorzik U, Wolf BA (1997) Reduction of the interfacial tension between poly(dimethylsiloxane) and poly(ethylene oxide) by block copolymers: Effects of molecular architecture and chemical composition. Macromolecules 30(16):4713–4718
36. Khandpur AK et al. (1995) Compatibilizers for A/B blends: A-C-B triblock versus A-B diblock copolymers. Polyblends'95, SPE Regional Technical Conference on Polymer Alloys and Blends. Boucherville, Quebec, Oct 19–20, pp 88–96
37. Fayt R, Jerome R, Teyssie P (1989) Molecular design of multicomponent polymer systems. XIV Control of the mechanical properties of polyethylene-polystyrene blends by block copolymers. J Poly Sci Part B: Poly Phys 27(4):775–793
38. Epstein BN (1977) US Patent 4 172 859
39. Freluche M et al. (2006) Graft copolymers of poly(methyl methacrylate) and polyamide-6: Synthesis by reactive blending and characterization. Macromolecules 39:6905
40. Pernot H et al. (2002) Design and properties of co-continuous nanostructured polymers by reactive blending. Nat Mat 1:54
41. Charoensirisomboon P et al. (2000) Polymer 41:5977
42. Charoensirisomboon P et al. (1999) Polymer 40:6803
43. Charoensirisomboon P, Inoue T, Weber M (2000) Polymer 41:4483
44. Charoensirisomboon P, Inoue T, Weber M (2000) Polymer 41:6907
45. Orr CA et al. (1997) Flow-induced reactive self-assembly. Macromolecules 30(4):1243–1246
46. Yin Z et al. (2001) Macromolecules 34:5132
47. Fleischer CA, Morales AR, Koberstein JT (1994) Interfacial modification through end group complexation in polymer blends. Macromolecules 27(2):379–85
48. Moskala EJ et al. (1984) On the role of intermolecular hydrogen bonding in miscible polymer blends. Macromolecules 17(9):1671–1678
49. Maric M, Ashurov N, Macosko CW (2001) Reactive blending of poly(dimethyl siloxane) with nylon 6 and polystyrene: effect of reactivity on morphology. Poly Eng Sci 41(4):631–642
50. Boutevin B, Guida-Pietrasanta F, Ratsimihety A (2000) Side group modified polysiloxanes. In: Silicon-Containing Polymers. Kluwer, Rotterdam, pp 79–112
51. Mohanty S, Santra RN, Nando GB (1997) Reactive blending of ethylene-methyl acrylate copolymer and poly-dimethyl siloxane rubber: kinetics studies from infrared spectroscopy. Adv Poly Technol 16(4):323–329
52. Santra RN et al. (1993) Thermogravimetric studies on miscible blends of ethylene-methyl acrylate copolymer (EMA) and polydimethylsiloxane rubber (PDMS). Thermochim Acta 219(1–2):283–292
53. Bhattacharya AK et al. (1995) Studies on miscibility of blends of poly(ethylene-co-methyl acrylate) and poly(dimethyl siloxane) rubber by melt rheology. J Appl Poly Sci 55(13):1747–1755
54. Jana RN, Nando GB (2003) Chemorheological study of compatibilized blends of low-density polyethylene and polydimethyl siloxane rubber. J Appl Poly Sci 88(12):2810–2817
55. Jana RN, Nando GB (2003) Thermogravimetric analysis of blends of low-density polyethylene and poly(dimethyl siloxane) rubber: The effects of compatibilizers. J Appl Poly Sci 90(3):635–642

56. Santra RN et al. (1993) In-situ compatibilization of low-density polyethylene and polydimethylsiloxane rubber blends using ethylene-methyl acrylate copolymer as a chemical compatibilizer. J Appl Poly Sci 49(7):1145–1158
57. Santra RN et al. (1995) In-situ compatibilization of thermoplastic polyurethane and polydimethyl siloxane rubber by using ethylene methyl acrylate copolymer as a reactive polymeric compatibilizer. Adv Poly Technol 14(1):59–66
58. Jana RN, Mukunda PG, Nando GB (2003) Thermogravimetric analysis of compatibilized blends of low density polyethylene and poly(dimethyl siloxane) rubber. Poly Degrad Stabil 80(1):75–82
59. Munoz PMP et al. (2001) High-density polyethylene modified by polydimethyl siloxane. J Appl Poly Sci 82(14):3460–3467
60. Shih W-C et al. (1999) Polydimethylsiloxane containing isocyanate group-modified epoxy resin: curing, characterization, and properties. J Appl Poly Sci 73(13):2739–2747
61. Scott HG (1972) Crosslinking of olefinic polymers and copolymers. US Patent 3 646 155
62. Shieh Y-T, Tsai T-H (1998) Silane grafting reactions of low-density polyethylene. J Appl Poly Sci 69(2):255–261
63. Swarbrick PGWJ, Maillefer C (1978) Manufacture of extruded products. US Patent 4 117 195
64. Chorvath I et al. (2002) Polyolefin thermoplastic silicone elastomers employing radical cure. WO Patent 2 002 088 247
65. Jana RN, Bhunia HP, Nando GB (1997) An investigation into the mechanical properties and curing kinetics of blends of low-density polyethylene and poly(dimethylsiloxane) rubber. Thermochim Acta 302(1–2):1–9
66. Kole S, Roy S, Bhowmick AK (1994) Interaction between silicone and EPDM rubbers through functionalization and its effect on properties of the blend. Polymer 35(16):3423–3246
67. Kole S, Roy S, Bhowmick AK (1995) Influence of chemical interaction on the properties of silicone-EPDM rubber blend. Polymer 36(17):3273–3277
68. Badesha SS et al. (2000) Compatibilized blend of fluoroelastomer and polysiloxane useful for printing machine component. Xerox, Stamford, CT, USA, p 6
69. Chorvath I et al. (2004) Fluoroplastic silicone vulcanizates. WO Patent 2 004 108 822
70. Furukawa H, Nakamura A, Shirahata A (1996) Preparation of siloxane-thermoplastic resin compositions with reduced surface siloxane bleed. US Patent 5 604 288
71. Jalali-Arani A, Katbab AA, Nazockdast H (2003) Preparation of thermoplastic elastomers based on silicone rubber and polyethylene by thermomechanical reactive blending: Effects of polyethylene structural parameters. J Appl Poly Sci 90(12):3402–3408
72. Shen J, Ye N (2001) Study on reaction kinetics of silane grafted HDPE and LLDP. Hecheng E Shuzhi Ji Suliao 18(3):9–12
73. Kole S et al. (1995) Grafting of silicone rubber onto polypropylene or polyethylene. Polym Networks Blend 5(3):117–122
74. Brook MA (2000) Silicones. In: Matison BMJ (ed) Silicon in Organic Organometallic and Polymer Chemistry. Wiley, New York, pp 256–308
75. Hamurcu EE, Baysal BM (1993) Interpenetrating polymer networks of poly(dimethylsiloxane): 1. Preparation and characterization. Polymer 34(24):5163–5167
76. Turner J, Cheng Y-L (2001) Process for preparing interpenetrating polymer networks of controlled morphology. US Patent 6 331 578

77. Turner JS, Cheng YL (2000) Preparation of PDMS-PMAA Interpenetrating polymer network membranes using the monomer immersion method. Macromolecules 33(10):3714–3718
78. Turner JS, Cheng YL (2004) pH dependence of PDMS-PMAA IPN morphology and transport properties. J Membr Sci 240(1–2):19–24
79. Turner JS, Cheng YL (2003) Morphology of PDMS-PMAA IPN membranes. Macromolecules 36(6):1962–1966
80. Robert C, Bunel C, Vairon J-P (1995) Hydrophilic, transparent material with high oxygen permeability containing interpenetrating polymer networks for use in soft contact lenses. EP Patent 643 083
81. Abbasi F, Mirzadeh H, Katbab AA (2002) Sequential interpenetrating polymer networks of poly(2-hydroxyethyl methacrylate) and polydimethylsiloxane. J Appl Poly Sci 85(9):1825–1831
82. Hron P et al. (1997) Silicone rubber-hydrogel composites as polymeric biomaterials. IX Composites containing powdery polyacrylamide hydrogel. Biomaterials 18(15):1069–1073
83. Lopour P, Janatova V (1995) Silicone rubber-hydrogel composites as polymeric biomaterials. VI Transport properties in the water-swollen state. Biomaterials 16(8):633–640
84. Duckova K et al. (1993) Silicone rubber-hydrogel composites as polymeric biomaterials. Part 5. Transdermal therapeutic systems based on hydrogel-filled silicone rubber. Eur J Pharm Biopharm 39(5):208–211
85. Lopour P et al. (1993) Silicone rubber-hydrogel composites as polymeric biomaterials. IV Silicone matrix-hydrogel filler interaction and mechanical properties. Biomaterials 14(14):1051–1055
86. Lednicky F et al. (1991) Silicone rubber-hydrogel composites as polymeric biomaterials. III An investigation of phase distribution by scanning electron microscopy. Biomaterials 12(9):848–852
87. Cifkova I et al. (1990) Silicone rubber-hydrogel composites as polymeric biomaterials. I Biological properties of the silicone rubber-p(HEMA) composite. Biomaterials 11(6):393–396
88. Lopour P et al. (1990) Silicone rubber-hydrogel composites as polymeric biomaterials. II Hydrophilicity, permeability to water-soluble low-molecular-weight compounds. Biomaterials 11(6):397–402
89. Sulc J, Vondracek P, Lopour P (1986) Hydrophilic silicone composites. DE Patent 3 616 883
90. Abbasi F, Mirzadeh H, Katbab AA (2002) Comparison of viscoelastic properties of polydimethylsiloxane/poly(2-hydroxyethyl methacrylate) IPNs with their physical blends. J Appl Poly Sci 86(14):3480–3485
91. Falcetta JJ, Friends GD, Niu GCC (1975) Molding from an interpenetrating network polymer. DE Patent 2 518 904
92. Huang G-S, Li Q, Jiang L-X (2002) Structure and damping properties of polydimethylsiloxane and polymethacrylate sequential interpenetrating polymer networks. J Appl Poly Sci 85(3):545–551
93. He X et al. (1995) Preparation of interpenetrating acrylic polymer-siloxane networks. US Patent 5 424 375
94. He XW et al. (1989) Poly(dimethylsiloxane)/poly(methyl methacrylate) interpenetrating polymer networks: 1. Efficiency of stannous octoate as catalyst in the formation of poly(dimethylsiloxane) networks in methyl methacrylate. Polymer 30(2):364–368

95. He XW et al. (1992) Poly(dimethylsiloxane)/poly(methyl methacrylate) interpenetrating polymer networks. 2. Synthesis and properties. Polymer 33(4):866–871
96. Caille JR (2000) Macromol Symp 153:161–166
97. Brachais L et al. (2002) Solid-state organization of poly(methyl methacrylate)-poly(methylphenylsiloxane) based interpenetrating networks. Polymer 43(6):1829–1834
98. Bischoff RA et al. (1998) Interpenetrating polysiloxane-polymethacrylate networks. FR Patent 2 757 528
99. Miyata T et al. (1996) Preparation of polydimethylsiloxane/polystyrene interpenetrating polymer network membranes and permeation of aqueous ethanol solutions through the membranes by pervaporation. J Appl Poly Sci 61(8):1315–1324
100. Tsumura M, Iwahara T (2000) Silicon-based materials prepared by IPN formation and their properties. J Appl Poly Sci 78(4):724–731
101. Fichet O et al. (2005) Polydimethylsiloxane-cellulose acetate butyrate IPN synthesis and kinetic study, Part I. Polymer 46:37–47
102. Darras V et al. (2004) Novel single and interpenetrating networks based on fluorinated polysiloxanes. Abstracts of Papers, 227th ACS National Meeting, Anaheim, CA, March 28–April 1 2004, p POLY-513
103. Frisch HL, Gebreyes K, Frisch KC (1988) Synthesis and characterization of semi- and full-interpenetrating polymer networks of poly(2,6-dimethyl-1,4-phenylene oxide) and polydimethylsiloxane. J Poly Sci Part A: Poly Chem 26(9):2589–2596
104. Gebreyes K, Frisch HL (1988) Improved synthesis and characterization of interpenetrating polymer networks of poly(2,6-dimethyl-1,4-phenylene oxide) (PPO) and poly(dimethylsiloxane) (PDMS). J Poly Sci Part A: Poly Chem 26(12):3391–3395
105. Frisch HLHMW (1991) Birefringence in Interpenetrating Polymer Networks of Poly(2,6-Dimethyl-1,4-Phenylene Dioxide)/Polydimethylsiloxane. J Poly Sci Part A: Poly Chem 29:131–133
106. Ebdon JR, Hourston DJ, Klein PG (1984) Polyurethane-polysiloxane interpenetrating polymer networks. 1. A polyether urethane-poly(dimethylsiloxane) system. Polymer 25(11):1633–1639
107. Ebdon JR, Hourston DJ, Klein PG (1986) Polyurethane-polysiloxane interpenetrating polymer networks: 2. Morphological and dynamic mechanical studies. Polymer 27(11):1807–1814
108. Klein PG, Ebdon JR, Hourston DJ (1988) Polyurethane-polysiloxane interpenetrating networks: 3. Polyetherurethane-poly(phenylmethylsiloxane) systems. Polymer 29(6):1079–1085
109. Zhou P, Xu Q, Frisch HL (1994) Kinetics of simultaneous interpenetrating polymer networks of poly(dimethylsiloxane-urethane)/poly(methyl methacrylate) formation and studies of their phase morphology. Macromolecules 27(4):938–946
110. Zhou P et al. (1993) J Poly Sci Part A: Poly Chem 31:2481
111. Frisch KC, Frisch HL, Klempner D (1972) Reseaux de polymères entremêlés. FR Patent 2 110 159
112. Dobkowski Z, Zielecka M (2002) Thermal analysis of the poly(siloxane)-poly(tetrafluoroethylene) coating system. J Therm Analy Calorim 68(1):147–158
113. Jones AS et al. (2000) Amide-type polymer/silicone polymer blends and processes of making the same. WO Patent 2 000 078 842
114. Murray DL, Hale WR, Jones AS (2000) Nylon 6-silicone blends. WO Patent 2 000 078 845

115. Liao J, Shearer G, Gross C (2003) Silicone TPV offers high performance solutions. Rubber World 227(5):40–43
116. Arkles B, Carreno C (1984) Filler-free models for the role of organofunctional silanes in composites. Polymeric Mat Sci Eng 50:440–443
117. Marciniec B (1992) Comprehensive Handbook on Hydrosilylation. Pergamon Press, Oxford, UK
118. Arkles B, Crosby J (1990) Polysiloxane-thermoplastic interpenetrating polymer networks. Advances in Chemistry Series 224(Silicon-Based Polym Sci), pp 181–199
119. Arkles BC (1983) Thermoplastic-curable silicone blends. US Patent 4 500 688
120. Arkles B (1983) A reactive processing method for IPN thermoplastics. Polymeric Mat Sci Eng 49:6–9
121. Gornowicz G et al. (2003) Thermoplastic elastomers containing polyurethanes and silicone. WO Patent 2 003 035 757
122. Gornowicz GA et al. (2000) Thermoplastic silicone elastomers and their preparation. US Patent 6 013 715
123. Zolotnitsky M (1997) Composition and method for impact modification of thermoplastics. US Patent 5 648 426
124. Brewer C et al. (2003) Thermoplastic silicone elastomers from compatibilized polyamide resins. WO Patent 2 003 035 759
125. Brewer CM et al. (2002) Thermoplastic silicone elastomers from compatibilized polyamide resins. US Patent 2 002 091 205
126. Chorvath I et al. (2002) Thermoplastic elastomer compositions containing silicone rubber and nylon resins and the dynamically vulcanizing method. US Patent 2 002 086 937
127. Chorvath I et al. (2001) Vulcanized thermoplastic silicone elastomers from nylon resins and polysiloxanes. WO Patent 2 001 072 903
128. Chorvath I et al. (2001) Silicone rubber-toughened thermoplastic resin composition. WO Patent 2 001 018 116
129. Crosby JM, Hutchins MK (1986) Fiber-reinforced thermoplastics containing silicone interpenetrating polymer networks. EP Patent 194 350
130. Fournier FM, Rabe RL (2004) Polyamide based thermoplastic silicone elastomers. US Patent 2 004 014 888
131. Chorvath I et al. (2002) Polysiloxane-based thermoplastic rubber containing polyester resins. WO Patent 2 002 046 310
132. Ward SK, O'Brien (1989) Enhanced GS melt extrusion of thermoplastics containing silicone interpenetrating polymer networks. EP Patent 308 836
133. Gross C, Lee M, Liao J (2003) Thermoplastic silicone elastomers from compatibilized polyester resins. WO Patent 2 003 035 764
134. Romenesko DJ, Mullan SP (1993) Poly(phenylene ether) resin modified with silicone rubber powder. EP Patent 543 597
135. Gornowicz GA (2000) Thermoplastic elastomers based on fluorocarbon resins and silicones. US Patent 6 015 858
136. Chung JYJ, Mason JP (1996) Toughened aromatic polycarbonate containing silicone rubber powder as molding composition. US Patent 5 556 908
137. Mason JP et al. (1997) Impact-modified polyamide-based molding composition. US Patent 5 610 223
138. Bilgrien CJ et al. (1992) Storage-stable flowable organosiloxane composition powders and their preparation. US Patent 5 153 238
139. Romenesko DJ, Buch RR (1995) Silicone resin powder for improving fire retardancy of organic resins. US Patent 5 391 594

140. Fustin CA et al. (2002) Reactive blending of functional polysiloxanes with poly(butylene terephthalate): clarification of reaction mechanisms and kinetics from a model compound study. J Poly Sci Part A: Poly Chem 40(12):1952–1961
141. Itoh K, Fukuda T (1978) Thermally curable silicone rubber compositions. US Patent 4 164 491
142. Chorvath I et al. (2002) Thermoplastic silicone elastomers employing radical cure. US Patent 6 465 552
143. Fu FS, Mark JE (1988) Elastomer reinforcement from a glassy polymer polymerized in-situ. J Poly Sci Part B: Poly Phys 26(11):2229–2235
144. Liang YF (1987) Arylene sulfide polymers of improved impact strength. US Patent 4 708 983
145. Liang YF, Beever WH (1985) Rubbery compounds as modifiers for poly(arylene sulfide). US Patent 4 888 390
146. Fujiki M, Furuta D, Naito M (2004) Manufacture of semi-IPN (interpenetrating polymer network) composite and the composite made of crosslinkable siloxane and radically polymerized polymer. JP Patent 2 004 263 062
147. Arkles BC, Smith RA (1990) Secondary crosslinked siloxane semiinterpenetrating polymer networks and methods of making them. US Patent 4 970 263
148. Gilmer TC et al. (1996) Synthesis, characterization, and mechanical properties of PMMA/poly(aromatic/aliphatic siloxane) semi-interpenetrating polymer networks. J Poly Sci Part A: Poly Chem 34(6):1025–1037
149. Gornowicz GA, Chang HS (2000) Thermoplastic silicone vulcanizates prepared by condensation cure. US Patent 6 153 691
150. Knaub P, Camberlin Y (1988) Gerard JF, New reactive polymer blends based on poly(urethane ureas) (PUR) and polydisperse poly(dimethylsiloxane) (PDMS): control of morphology using a PUR-b-PDMS block copolymer. Polymer 29(8):1365–1377
151. Vlad S, Vlad A, Oprea S (2002) Interpenetrating polymer networks based on polyurethane and polysiloxane. Eur Poly J 38(4):829–835
152. Xiao H et al. (1990) The synthesis and morphology of semi-interpenetrating polymer networks based on polyurethane-poly(dimethylsiloxane) system. J Poly Sci Part A: Poly Chem 28(3):585–594
153. Morin A (1990) Thermoplastic polycondensate-silicone blends and their preparation. FR Patent 2 640 632
154. Arkles BC (1987) Manufacture of curable silicone semi-interpenetrating networks. US Patent 4 714 739
155. Kohara S et al. (1999) Thermoplastic olefin elastomer compositions with low compression set, no oil bleeding, and good weather resistance. JP Patent 11 181 172
156. Sibahara S, Sugisaki A, Iwasa T (2000) Thermoplastic olefin rubber compositions for oil- and weather-resistant moldings. WO Patent 2 000 043 447
157. Medsker RE et al. (2000) Process for silicon hydride curing of thermoplastic vulcanizates. US Patent 6 150 464

Editors: Martin Möller · Helmut Keul

Functional Materials Derived from DNA

Xiang Dong Liu[1] · Masanori Yamada[2] · Masaji Matsunaga[1] ·
Norio Nishi[1,3] (✉)

[1]Nissei Bio Co., Ltd, Megumino, Eniwa, 061-1374 Hokkaido, Japan
nishin@ees.hokudai.ac.jp
[2]Faculty of Science, Okayama University of Science, Ridaicho, 700-0005 Okayama, Japan
[3]Graduate School of Engineering, Hokkaido University, Kita-ku, 060-8628 Sapporo, Japan

1	Introduction	150
2	Helical Anionic Polymer	152
2.1	DNA and Inorganic Matters	153
2.2	DNA and Collagen	155
2.3	DNA and Cationic Lipids	156
3	DNA Strand Pairing	158
3.1	DNA Nanostructure	158
3.2	DNA-Mediated Molecular Structures	161
3.3	DNA Device	163
4	Double Helix Binding	165
5	Biopolymer	171
6	Perspective	174
References		175

Abstract DNA has special properties and its unique double-helical structure offers excellent prospects for creating novel DNA-based materials. In recent years, DNA has been shown to be an ideal molecule in the material world. This review is intended to provide an overview of functional materials derived from DNA based on the double-helical structure. Various DNA-based materials are reviewed according to the basic DNA structural properties, including the electrostatic properties of DNA as a highly charged polyelectrolyte, complementary base pairing, and intercalation and groove binding interaction with small molecules. Finally, attempts to produce biomaterials based on DNA are also summarized.

Keywords DNA binder · Double-helical structure · Intercalation · Molecule device · Nanostructure

Abbreviations
dsDNA Double-stranded DNA
ssDNA Single-stranded DNA

B-DNA	B form DNA
A-DNA	A form DNA
Z-DNA	Z form DNA
SEM	Scanning electron microscope
TEM	Transmission electron microscope
AFM	Atomic force microscopy
1D	One-dimensional
2D	Two-dimensional
3D	Three-dimensional
PCR	Polymerase chain reaction
CD	Circular dichroism
AuNPs	Gold nanoparticles
DX	Double crossover
PX	Paranemic crossover
TX	Triple crossover
PAHs	Polycyclic aromatic hydrocarbons
PCDD	Polychlorinated dibenzo-*p*-dioxin
PCDF	Polychlorinated dibenzofuran
PCB	Polychlorinated biphenyl
PSf	Polysulfone

1
Introduction

DNA is a unique molecule, having seemingly magical physical and chemical properties and an unusual double-helical structure. Investigation of DNA has had a profound impact on developments in the life sciences since 1953, when Watson and Crick proposed the double-helical structure. The particular properties of DNA have also attracted researchers in materials. After various explorations, in recent years DNA has been shown to be an ideal molecule in the material world. DNA possesses properties that offer excellent prospects for creating novel DNA-based materials. These properties include the electrostatic properties of DNA as a highly charged polyelectrolyte, the complementary base pairing, and the configuration characteristics induced intercalation and groove binding. All of these important properties are due to the double-helical structure of DNA. Herein, in order to provide an overview of the functional materials derived from DNA, various studies will be reviewed based on the associated structural property of DNA (Fig. 1).

Research on DNA-based materials also depends on the facility of obtaining various DNA samples. Profiting from the life sciences, the automated synthesis method is virtually able to synthesize designed DNA, and the polymerase chain reaction (PCR) can amplify the DNA sequence. On the other hand, as a native substance widely existing in organisms, DNA has a broad source in the natural world, especially, from fisheries. A typical case is the salmon milt, which is mostly treated as feedstuff. The DNA content in salmon milt is over

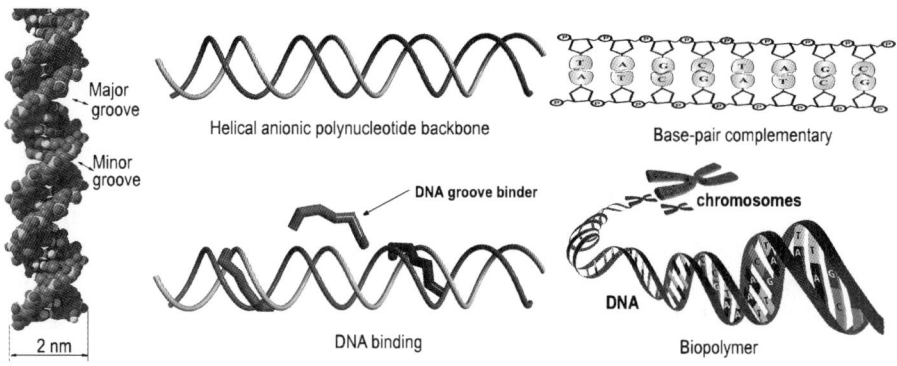

Fig. 1 Basic features of dsDNA

10% in dry weight, and the DNA can be extracted from salmon milt by a very simple process. World production of salmon has increased to over 2.4 million tons in 2004. Based on this catch amount, it is estimated that about 3000 tons of DNA can be obtained every year just from the salmon supply. Such facile sources of DNA effectively promoted research of DNA-based materials for use in further applications.

According to the Watson–Crick model, a DNA molecule consists of two polynucleotide strands coiled around each other in a helical fashion, which is often described as a double helix. Each strand has a backbone made up of sugar molecules linked together by phosphate groups. The orientation of the two strands is antiparallel. The backbone of double-stranded DNA (ds-DNA) is highly negatively charged. The two negatively charged backbones lead to DNA "stiffness", with a persistence length of around 50 nm, which corresponds to 150 base pairs in the double helix [1]. This property allows dsDNA to form highly ordered complexes with electrostatic binding agents such as metal ions, cationic surfactants, and polycationic agents. The reports selected relating to DNA electrostatic binding will be detailed in Sect. 2 on helical anionic polymers.

The opposite strands of dsDNA are held in precise register by regular base pairing between the two strands. This base-pair complementarity is a consequence of the size, shape, and chemical composition of the bases. The highly specific A-T and G-C Watson–Crick hydrogen bonding provides the chemical foundation for genetics. This structurally well-defined cohesive combining of DNA stands has been tried in nanotechnology to direct the assembly of highly structured materials with specific nanoscale features. Section 3 on DNA strand pairing will give an overview on DNA nanoarchitectures such as self-assemble structure, DNA arrays, and DNA nanomechanical devices.

The 3D structure of DNA is known in detail and this has permitted the meaningful structural and mechanistic interpretation of DNA–agent complexes. Two right-handed antiparallel double helix forms of DNA are well known, A- and B-DNA, and a third left-handed antiparallel duplex referred to as Z-DNA has been identified. The most common DNA structure is the B-DNA. The X-ray diffraction pattern of DNA indicates that the stacked bases are regularly spaced 0.34 nm along the helix axis. About ten pairs make a complete turn every 3.4 nm, with a 2 nm diameter. The B-DNA possesses a wide major groove and a narrow minor groove of approximately the same depth. On the other hand, in the DNA helix structure, both the sugar and phosphate (which constitute the backbone) are quite soluble in water, but the DNA bases (which are in the middle of the helix) are relatively hydrophobic and insoluble. Therefore, two adjacent flat base pairs form a local hydrophobic environment. It has been shown that a number of small molecules can bind directly and selectively to DNA. This specific interaction occurs either in the groove or by inserting the molecule between two base pairs (intercalation). Both patterns are allowed by the double-helical structure. In many instances, the small molecules are highly toxic or carcinogenic. In Sect. 4 on dsDNA binding, we will focus on applications based on the DNA groove binding and intercalation.

DNA is a biopolymer that widely exists in the natural world. DNA possesses perfect biophysical and biochemical properties, which have been optimized over billions of years of evolution. These unique properties of DNA offer excellent prospects for serving as a construction material in bioscience. Several attempts have been made to use DNA as a biomaterial. These publications will be summarized in Sect. 5 on biopolymers.

2
Helical Anionic Polymer

DNA is a helical, linear polymer of deoxyribonucleotides linked 3' to 5' by phosphodiester bridges. The double helix chain has a width of 2 nm and a length of 0.34 nm per nucleoside subunit. A wide range of chain lengths, from nanometers to tens of microns, can be realized with established technology in molecular biology, for example DNA ligation, enzymatic digestion, and PCR. For the B-DNA, negatively charged phosphate groups in the phosphate backbone are regularly spaced around the central axis of the double helix. These phosphate groups play very important roles in the living body. For example, positively charged histones interact via ionic bonds with the negatively charged phosphate groups on the polynucleotide backbone to form chromosomes, and several metal ions are involved in the DNA replication, repair, and transcription. In material science, the charged phosphate groups can conceivably be used to derive various functional materials with cations. Actually, during the last decade, many DNA materials associated with metal ions, cationic lipids or surfactants, cationic proteins, cationic nanoparticles, and even polycationic macromolecules have been reported. These achievements demonstrate the promising feasibility of using the anionic DNA molecule to create precisely ordered materials with various functions.

2.1
DNA and Inorganic Matters

In contrast to other familiar anionic polymers, the anionic dsDNA is distinguished by the double-helical conformation and the existence of the pairing bases. These characteristics make the interaction between the DNA molecule and various metal ions more complex. A recent investigation indicated that the character of the DNA–metal ion complex strongly depends on the ion type [2]. Al^{3+}, Cr^{3+}, Fe^{2+}, Fe^{3+}, Cu^{2+}, Zn^{2+}, Y^{3+}, La^{3+}, In^{2+}, and Cd^{2+} metal ions at 1 M concentration immediately form a water-insoluble gel from highly concentrated solutions of DNA (10 mg/mL) via a ion cross-linking. On the other hand, alkali earth metal ion solutions did not yield gels. The SEM images taken from the surface of the DNA–ion complex suggest that the structure of the DNA–Al^{3+} complex is more regular than the DNA–Cu^{2+} complex (Fig. 2). However, the results from CD and infrared spectra indicated that Cu^{2+} not only strongly binds to the phosphate group by electrostatic interaction but also changes the DNA base pairs, leading to destruction of the B-DNA structure and to a decrease of flexibility of the DNA molecule.

Since current top-down lithography techniques faced a serious problem for electronic technologies, interest in the bottom-up approach has risen sharply in recent years. Among many candidate materials for the bottom-up strategies, DNA attracted a lot of attention since it has a small diameter of about 2 nm, is flexible, and can be targeted uniquely at each end. One application for DNA being explored is "nanowires" for interconnecting quantum dots with each other and for linking them to macroscopic electrodes or other devices. Since a DNA molecule in aqueous solution is usually randomly structured as a result of thermal fluctuations, the individual DNA molecules must be separated and stretched prior to serving as templates for nanowire fabrication.

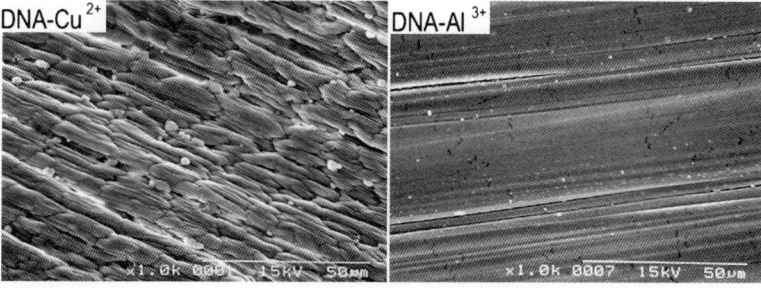

Fig. 2 SEM images of DNA–metal ion fiber of DNA–Cu^{2+} fiber and DNA–Al^{3+} complex. The structure of the DNA–Al^{3+} complex is more regular than that of DNA–Cu^{2+} complex

Many methods have been used to stretch and align DNA molecules, including molecular combing [3, 4], electrophoretic stretching [5, 6], and hydrodynamic stretching [7]. The simplest and most widely used protocol is molecular combing, schematically drawn in Fig. 3. In this method, DNA molecules are attached by their ends to a glass surface, while suspended in a droplet of solution. This droplet is then allowed to evaporate, causing the retreating meniscus to exert a traction on the molecule and resulting in its extension as it becomes bound to the glass surface.

Following stretching and positioning, the DNA molecule is generally metallized to enhance its conductivity. The most common method for metallizing DNA is electroless plating involving the reduction of metal ions electrostatically bound to the DNA backbone. For example, Braun et al. [8] prepared silver nanoparticles along the DNA to enhance the conductivity of the DNA wire. The schematic process is shown in the Fig. 4. After DNA stretching and positioning achieved by hydrodynamic flow, Ag ions were bound to the DNA molecule by an Ag – Na ion exchange treatment. Hydroquinone reduces DNA-bound Ag ions to Ag(0) metallic clusters, which then autocatalyze further reduction of Ag ions from solution. The AFM image (Fig. 4) shows an

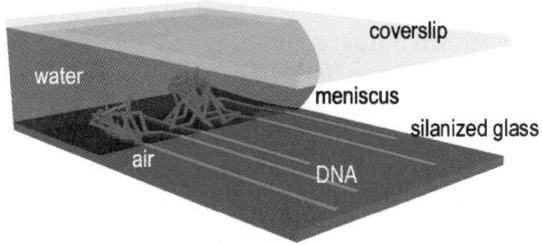

Fig. 3 Schematic mechanism of DNA molecular combing. The meniscus generates a surface tension during evaporation, which stretches DNA. Based on [4]

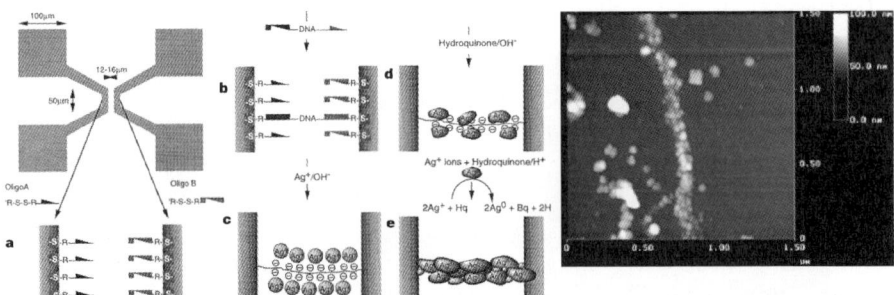

Fig. 4 Construction of a silver wire connecting two gold electrodes. The *left* images show the process in the experiments. The *right* photo is the AFM image. Note the granular morphology of the conductive wire. Reprinted with permission from [8]

Ag nanowire positioned between two Au electrodes. A subsequent publication reported not only making conducting gold and silver nanowires from DNA templates but also protecting specific regions of DNA molecules from metal deposition by associating proteins along sections of the DNA [9]. Similar to these two examples, various metallic nanowires prepared by the deposition of palladium [10], platinum [11, 12], copper [13] and cobalt [14] metal on DNA have been investigated as an approach for creating conductive nanowires.

Not only metal ions, but also positively charged nanoparticles have been templated onto DNA molecules. Ohtani and coworkers [15] prepared novel surface-functionalized AuNPs based on the conventional reduction of $HAuCl_4$ using aniline as a reducing agent. In their work, the AuNPs were successful assembled on DNA molecules by two different procedures. One was assembly of AuNPs on DNA that was previously aligned on the substrates. The other procedure first involved binding the AuNPs onto DNA and then subsequent stretching and fixation of the composite.

2.2
DNA and Collagen

The complex of DNA and collagen has potential uses for fabrication of new nanostructures. Collagen is the most important structural protein in the body. It acts as the "scaffolding" of the extracellular matrix and gives skin, tendon, cartilage, and intervertebral discs their mechanical strength. A collagen molecule consists of three polypeptide chains arranged in a parallel triple helix. These chains are unusually rich in glycine, proline, and hydroxyproline. They are held together by hydrogen bonds that link the peptide amine bonds of glycine residues to peptide carbonyl groups in an adjacent polypeptide. This results in a rigid rod-like triple helix geometry with a diameter of 1.5 nm and a length of about 300 nm for a single collagen molecule. Moreover, it was reported that there are locations of charged residues in the collagen structures [16]. As mentioned previously, the dsDNA molecule shows local stiffness in a range of about 100 nm but long-range flexibility of the double helix. Exploring the interaction between the two helical biopolymers is useful for building regular structures at the nanoscale.

The first morphological investigation about complex of dsDNA and collagen was carried out by TEM, and published in 1997 [17]. It was found that the pattern of the collagen fibrils formed in the presence of DNA was extremely regular compared with pepsin-digested collagen alone. In addition, the width of the fibrils was larger than that of the fibrils without DNA (Fig. 5a,b). This phenomena was confirmed again in a following publication [18]. The results showed that the presence of linear dsDNA molecules strongly promotes the fibril's formation, but the circular closed supercoil DNA molecules decelerate the complex formation. The phosphate groups of DNA are likely to

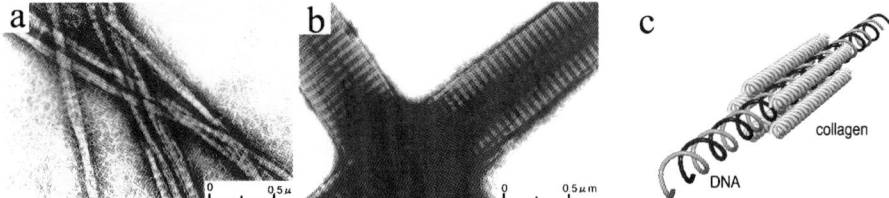

Fig. 5 TEM images of acid-soluble collagen fibrils formed **a** in the absence and **b** in the presence of DNA. **c** Schematic model of collagen–DNA complex formed in aqueous solution

interact with the charges in collagen and play key roles in the collagen fibril forming process. It has also been reported that some materials containing phosphate groups, such as phosphoproteins, promote collagen fibril forming process [19]. Recently, Mrevlishvili's group proposed a model of the complex of DNA and collagen [20, 21]. Their model proposes that the DNA helical structure becomes more rigid when the rigid rod-like structures of triple helix collagen (300 nm) overlap and envelop the backbone of DNA (Fig. 5c). Although only a few reports on the collagen–DNA complex have been published, these initial approaches are encouraging, as collagen-DNA complexes will open up future opportunities for the construction of nanoscale ordered materials.

2.3
DNA and Cationic Lipids

Cationic lipid or surfactant binding to DNA molecules can lead to compact, ordered complexes. In a series of works [22–27], Okahata et al. described the preparation and characterizations of the DNA–cationic lipid complexes (Fig. 6a). Generally, the DNA–lipid complex can be readily prepared by mixing aqueous solutions of anionic DNA and cationic lipid, which forms a water-insoluble precipitate. These complexes are often soluble in common organic solvents, and can be cast to form thin films. The results from the investigation of CD spectra of DNA strands showed that the self-assembly of DNA and the cationic lipids form a double-helical structure similar to B-DNA. The results from the polarized absorption spectra of intercalated dye molecules indicated that the DNA–lipid complexes exhibit intercalation behavior in chloroform solution. Interestingly, when the DNA–lipid cast film was stretched to three times in length, DNA strands could be aligned in the stretching direction [24]. X-ray diffraction of the cast film confirmed the ordered structure. In addition, as shown in Fig. 7, electric conductivity in the aligned DNA cast film was demonstrated [28, 29]. Kawabe [30–32] and coworkers found that some dye molecules intercalated in DNA–lipid complex films can lead to amplified spontaneous emission during nanosecond optical pumping. The durability

Functional Materials Derived from DNA

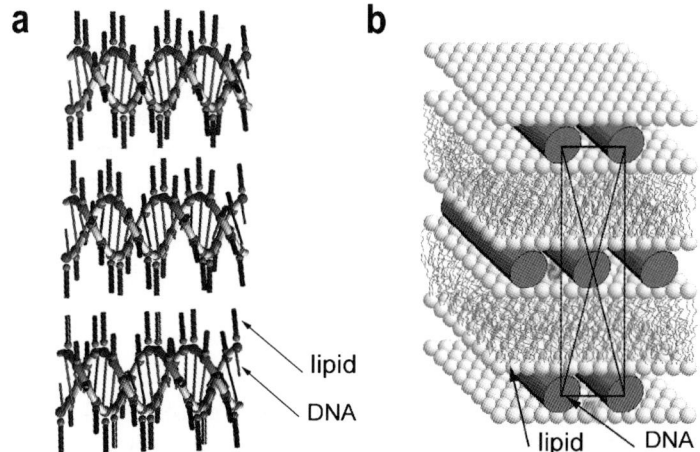

Fig. 6 Schematic representation of **a** complex of DNA–cationic lipid, and **b** DNA and a mixture of cationic lipid and neutral lipid, the lipid–DNA complexes intercalated in multilamellar membranes of cationic lipid and neutral lipid. Part **b** was reproduced with permission from [39]

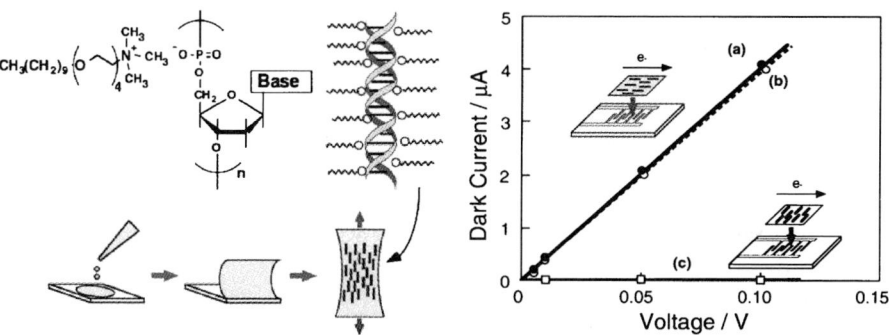

Fig. 7 Dark currents for aligned DNA–lipid complex films measured under three conditions: DNA strands are perpendicular to the two electrodes in atmosphere (*a*) and in a vacuum at 0.1 mmHg (*b*), contrasting with DNA strands parallel to the two electrodes (*c*). Reproduced with permission from [28]

and low threshold values suggest that the DNA–lipid complexes are practical candidates for thin-film dye lasers.

Another use of the DNA–lipid complex was reported by Wong and coworkers [33]. The DNA molecule structural features can be imprinted into CdS nanostructures by using self-assembled DNA-membrane templates. The initial application of the DNA–lipid multilamellar structure was the use as gene carriers in gene therapy, and many reports [34–39] were published (Fig. 6b). To prepare the DNA templates, anionic DNA was reacted with a mixture of cationic lipid and neutral lipid in solution, forming lamellar structures in which DNA chains are inserted between stacked lipid sheets. The Cd^{2+} ions were organized along the DNA chains in the multilamellar structure, and subsequently reacted with H_2S to form CdS nanorods. In addition, the widths of the nanorods can be tailored by adjusting the lipid mixture. This technique may lead to custom-designed crystals with useful electronic, magnetic, or optical properties.

3 DNA Strand Pairing

From an energetic point of view, the most important contribution to the DNA double-helical structure is the pairing bases between the two strands. Two interactions, hydrogen bonding between paired nucleobases and stacking interactions between adjacent nucleobases play important roles in the stability of the duplexes [40]. Combining the hydrogen bonds and the hydrophobic stacking interactions, one DNA strand can associate with and recognize other DNA stands, which leads to the Watson–Crick type right-handed helix. The specific recognition between complementary stretches of nucleotides makes it possible to encode instructions for assembly in a predetermined fashion on the nanometer scale. Besides the nanoscale structures made of DNA only, nanostructures made of DNA and other components, or mainly of other components have been attempted in various ways. The presence of DNA, with its controllable molecular properties, enables the creation of constructions with a predictable ordered structure. Moreover, based on an ingenious complementarily design, various reports demonstrated that the DNA strand stacking can also drive nanoscale movements.

3.1 DNA Nanostructure

The familiar dsDNA is a linear molecule not suitable for forming complex motifs. To build complex DNA nanoarchitectures, branched DNA molecules are needed. Fortunately, this key problem was resolved. By designing appropriate DNA sequences, the branched DNA molecules can be produced by the conventional solid support synthesis.

Generally, two types of the branched DNA molecule are used to generate complex DNA nanoarchitectures, which are named "junction" and "crossover", as illustrated in Fig. 8 [41, 42].

Early works on the artificial DNA architectures mainly used the branched junction motifs [43, 44]. Such junctions contain three or four arms of ssDNA oligomer (Fig. 8 motifs 1 and 2). Figure 9 shows four units of the branched junction assemble to produce a quadrilateral (motif 5). The outside

Functional Materials Derived from DNA 159

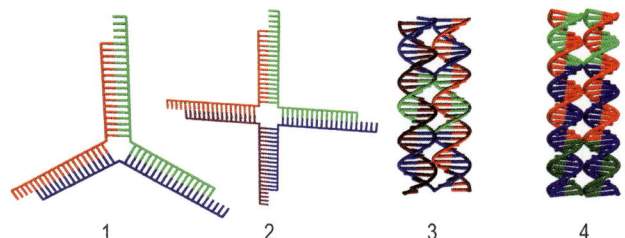

Fig. 8 Schematic representation of DNA junctions and crossover tiles. Motif 1 is a branched DNA junction with three arms and motif 2 with four arms. Every terminal in the arm is an unpaired ssDNA. The ssDNA acts as "sticky ends", which may pair with another complementary strand. The two motifs 3 and 4 are two different antiparallel double-crossover molecules containing an even number of half-helical turns between branch points (DAE) or an odd number (DAO). They are more stable and thus usually applied. Oligonucleotide strands are individually represented with different colors

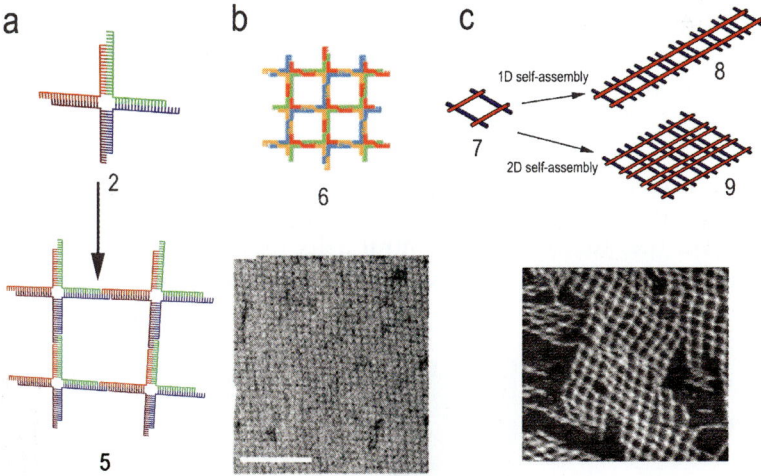

Fig. 9 Assembly of DNA junctions. **a** Four of the junctions in motif *2* are complexed to yield the structure in motif *5*. The complex has maintained open valences so that it could be extended by the addition of more monomers. **b** Square lattice formed from four-arm junctions held in a square-planar configuration (*6*) by protein RuvA, with TEM image of the lattice shown beneath. The *scale bar* represents 100 nm. Reprinted with permission from [47]. **c** 1D self-assembly of the motif *7* derives into a railroad track-like array *8*, and the 2D self-assembly produces a lattice array *9*. An AFM image of array *9* is shown beneath with a scan size of 400×400 nm^2. Adapted with permission from [45, 46]

of this quadrilateral contains further sticky ends so, in principle, this arrangement could be extended to produce a lattice in two or three dimensions. As shown in motif 7 in Fig. 9, such junctions form a rhomboidal supramolecular structure [45, 46]. A set of rhomboids was subsequently assembled into 1D "railroad tracks" (motif 8 in Fig. 9) as well as into 2D lattices (motif 9 in

Fig. 9) [45, 46]. Although the schematic representations of motif 2 and 5 in Fig. 9 may suggest the helices of a four-armed junction cross perpendicularly, they actually possess an angle of about 60°. AFM analyses showed that the torsion angles between helices are relatively constant throughout the entire lattice (Fig. 9) [45, 46]. By incorporating a protein (RuvA), a square-planar configuration (motif 6 in Fig. 9) has been built [47].

Subsequent works found that these simple junction motifs often did not yield desired regular superstructures because of their high conformational flexibility [48]. To solve this problem, a more rigid component called a "crossover tile" was developed. The simplest crossover tile is the double crossover (DX). DX tiles consist of two double-stranded helices, which interchange single strands at two crossover points [42]. Because of the decreased Coulomb repulsion, the two antiparallel motifs DAE and DAO, where the minor groove of one helix lies in the major groove of the other helix, are more stable (Fig. 8, motifs 3 and 4). Both the DAE and DAO have a rigidity comparable to linear duplex DNA [49], which makes them ideal building blocks for the construction of DNA-based materials with designable structures. Paranemic crossover (PX) molecules are composed of two parallel helices, and form crossovers at every point possible. This renders the tiles more stable than the comparable DX molecule [50, 51].

Based on these DX tiles, a number of other useful motifs were extended in this way. Triple crossover (TX) tiles can be regarded as being built from two DAO-DX tiles that share the central helix [52]. In the same way that DX tiles extend to the TX, more complex crossover tiles were developed. DNA

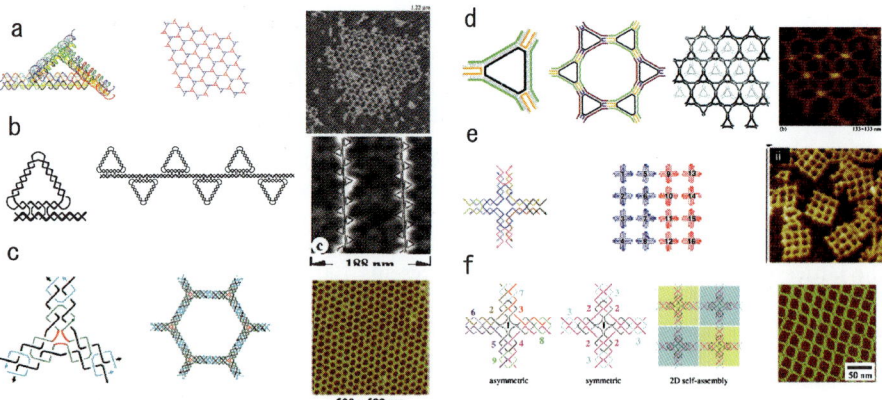

Fig. 10 Complex DNA motifs. **a** DX triangle self-assembly to a pseudohexagonal lattice [57]. **b** DNA triangles ligated to produce a linear array [58]. **c** DNA three-point star motif assembly to the hexagonal arrays [59]. **d** Hexagonal structure composed of six triangular complexes, and extended to a pair of overlapping hexagonal tilings [60]. **e** 16 cross-tiles construct directly to one square [62]. **f** Self-assembly of the cross motifs to 2D lattice [61]. Reproduced with permission from cited references

motifs containing two [53], three [54], four [53, 55], six [56], eight [55], and 12 [55] dsDNA domains have been reported. Such crossover tiles can be used as building blocks for the generation of larger DNA nanostructures. All of these motifs have found application in structural DNA nanotechnology.

However, the crossover tiles mentioned above are designed as DNA molecules crossing in their helices parallel and coplanar. Their lattices tend to grow very well in the direction parallel to the helix axes but have fairly poor variety in directions cross to the helix axes. In order to eliminate this problem, more complex crossover tiles with larger motifs such as triangles [57–60] or cross tiles [61–64] have been developed. Such motifs can be considered as flat and rigid building blocks to build the final desired superstructure. Some examples and their self-assembled products are shown in Fig. 10. More recently, controlled folding of a long single DNA strand was used to create an octahedron [65] and other arbitrary 2D shapes [66]. The authors demonstrated that this method could readily provide access to approximate structure outlines of any desired shape.

3.2
DNA-Mediated Molecular Structures

The base-pairing mechanism has been designed to direct the assembly of other functional molecules by the use of appropriate attachment chemistries. These are often hybrids of DNA and other compounds, such as small organic molecules, proteins, colloids or even AuNPs. In these materials, DNA is mainly used as a structuring element, which directs the assembly of subjects to self-assemble in highly ordered materials. Mainly, two different strategies have been applied: one uses the DNA molecule as a linker to connect other materials, and the other introduces other materials into the unit of the nanoscale building block of DNA nanostructure. The former is simple and convenient. After the synthesis of properly designed units, they self-assemble into a molecule–oligonucleotide hybrid, just by mixing the components in the right stoichiometry [67]. The latter has the ability to control the periodicity and interparticle spacings of the nanoarrays by design. However, for both methods, it is necessary to modify the non-DNA compounds with ssDNA.

Metal nanoparticles, especially AuNPs, are very familiar partners for forming DNA-mediated materials. Here, we will focus on the metal nanoparticles that have DNA oligomer ligands on the surface for possible organization, which are called "oligofunctional DNA–metal nanoparticles". Such oligofunctional DNA–metal nanoparticles can be readily prepared. DNA sequences up to about 120 nucleotides in length, modified with a large variety of chemical substituents, such as amino and thiol groups attached to the $3'$- or $5'$-terminus, are readily available by a multitude of commercial suppliers at relatively low cost [68]. The attachment of DNA oligomers on the AuNPs can be readily obtained by mixing citrate-stabilized AuNPs and alkylthiol derivatized DNA oligomers, and subsequent incubation and purification [69].

Alivisatos and coworkers showed for the first time that a discrete number of gold nanocrystals with one ssDNA strand ligand can be organized into spatially defined structure by base pairing [70]. Oligonucleotides modified at either the 3'- or 5'-terminus with a free sulforyl group were coupled with an excess of nanoparticles. After being combined with suitable ssDNA templates, parallel (head-to-tail) and antiparallel dimers (head-to-head) were obtained (Fig. 11a). UV/vis absorbance measurements indicated changes in the spectral properties of the nanoparticles as a consequence of the supramolecular organization [71]. Following this approach, Mao et al. reported a new challenge to build larger arrays by DNA-encoded self-assembly, which assembled AuNPs into 4 μm-long ssDNA strand templates produced from "rolling-circle DNA polymerization" [72]. In the rolling-circle DNA polymerization, a DNA polymerase uses a short (less than 100 bases long), circularized, single DNA strand as a template to synthesize long (more than 10000 bases), linear, tandemly repetitive single DNA strands under isothermal conditions [73]. The 5 nm gold nanoparticls modified by ssDNA strands with a thiol group at 5' end were introduced into the resulting long ssDNA strand.

To arrange AuNPs into a monolayer, Simon and coworkers [74] reported a simple protocol via the oligonucleotides complementary immobilization. 5'-Amino-modified oligonucleotide was immobilized on the substrate surface first, and then, 15 nm gold particles modified with thiolated DNA oligomers were coupled to the surface through DNA base pairing. The resulted nanoparticle monolayers were demonstrated to have a thermally activated conductivity. A similar approach has been used to insert a liposome into a DNA chip [75] (Fig. 11b).

Three-dimensionally linked nanoparticle assemblies have been successful prepared by DNA hybridization. Mirkin's group initially described a method

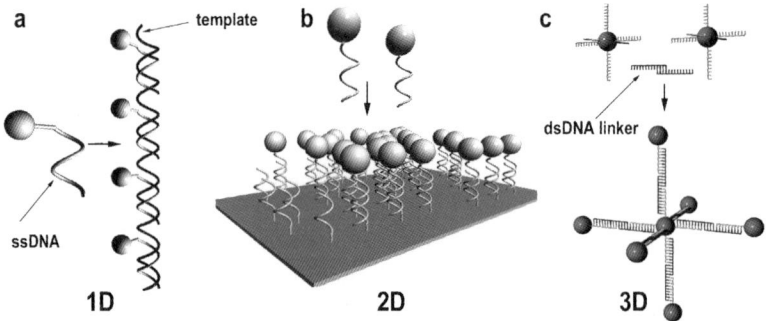

Fig. 11 Process of the DNA-based colloidal nanoparticle assembly. **a** Base-pairing interactions induced assembly in 1D template. **b** Immobilization by DNA hybridization onto 2D surface. **c** 3D assembly by duplex DNA interconnects

for assembling 13 nm AuNPs rationally and reversibly into macroscopic aggregates [76]. As shown in Fig. 11c, two non-complementary oligonucleotides are separately coupled to gold particles by thiol adsorption. Subsequently, as a linker, a DNA duplex with "sticky ends" that are complementary to the two oligonucleotides on the particles was added to the mixture. Then, the oligonucleotide-modified nanoparticles self-assemble into aggregates (Fig. 11c). This process is reversible by thermal denaturation. Following this initial work, the 3D self-assembly through complemented hybridization of DNA has been used to aggregate other particles such as polystyrene latex microspheres [77], micron-sized colloids [78, 79], and protein-encapsulated iron oxide (ferritin) [80].

DNA templated protein arrays with predictable control at the nanometer scale could lead to single-molecule detection in proteomics studies. Individual proteins placed at unique locations on the nanoarray could be detected with single molecule imaging techniques such as recognition imaging, in which specific antibodies are attached to the scanning probe cantilever.

Aptamers are DNA or RNA molecules that can bind other molecules such as other nucleic acids, proteins, small organic compounds, and even entire organisms [81]. Yan et al. incorporated aptamer sequences into a rationally designed DNA nanostructure, and successfully used the aptamer-bearing DNA nanostructure for the directed assembly of thrombin protein arrays [82]. They succeed in the use of a TX DNA tile as the template to direct the assembly of an aptamer and its subsequent organization of proteins into periodic 1D arrays. AFM images show that aptamers on the DNA array are mostly occupied by thrombin. The results clearly demonstrated that the thrombin binding aptamer still functions as the protein-binding moiety upon incorporation into a complex DNA nanostructure. In similar works [83, 84], the ability to use self-assembled DNA nanostructures to precisely control the spatial location of both streptavidin protein molecules and their nanogold conjugates has also been demonstrated.

Furthermore, by constructing a family of DNA tiles known as four-by-four DNA tiles, a precise control of periodic spacing between individual protein molecules was successfully demonstrated on two types of DNA nanoassemblies, the 2D nanogrid and the 1D nanotrack [85].

3.3
DNA Device

The DNA strand stacking is not only available to build supramolecular structures, it is also able to drive nanoscale movements. The first DNA nanoscale device is a simple example that drives a molecular device by changing the concentration of a small molecule [86]. This device consists of two DX molecules connected by a shaft with a special sequence that could be converted from the normal right-handed B-DNA to the left-handed Z-DNA. The transition can be triggered by the addition of cobalt hexamine ($[Co(NH_3)_6]^{3+}$). This conformational change results in a half-turn rotation of the two DX sections. Other approaches, such as pH [87–89] and temperature [90] changes, have also been reported.

Applying short ssDNA strands that selectively interact with a specific part of a molecular device is another method to make the device move. A robust rotary device was developed based on multiple crossover motifs [91]. The motivity of the device was the reversible binding of DNA strands. The central axis consists of a couple of ssDNA sections. The conformation of the two ssDNA strands can readily be switched between a PX conformation and its topoisomer conformation. The ssDNA strand replacement can cause the interconversion between the two conformations, and results in an 180° rotation of the end of one strand. This work showed that a rotary nanomechanical device is capable of being cycled by the addition of strands that direct its structure (Fig. 12a). As an application of the DNA nanodevice, a unique DNA nanomechanical device that enables the positional synthesis of products whose sequences are determined by the state of the device has been

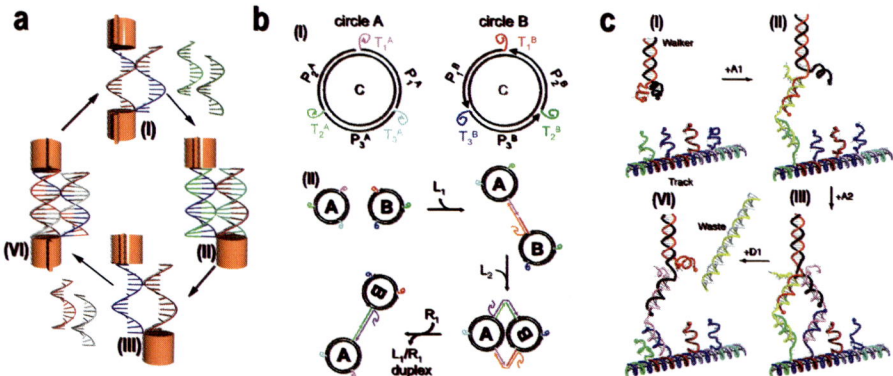

Fig. 12 DNA nanomechanical devices. **a** Scheme of the rotational device controlled by DNA hybridization. Insertion of the ssDNA strands leads to the formation of structure *I*. Structure *I* can be switched back to structure *III* by another branch-migration and strand-insertion process. **b** Scheme of the rolling process of gears. *I* Structures of the individual gears. *C* and *P* indicate DNA strands, and *T* indicates teeth. *II* Operation of the gears. *L* and *R* represent linker and removal strands, respectively. *L1* and *R1* are complementary to each other. Both circles remain intact during the rolling process. The only changed strands are the linker and removal strands. Note that no twisting motion will be generated to the central stands during the rolling process. Reprinted with permission from [93]. **c** DNA walker based on hybridization. *I* The walker consists of duplex DNA with two single-stranded "feet". The track is duplex DNA with single-stranded extensions as binding sites for the walker. *II* Walker attached to branch *A1*. *III* Walker attached to branches *A1* and *A2*. *IV*) Walker released from branch *A1* to yield duplex waste. Reprinted with permission from [94]

reported [92]. Like the translational capabilities of the ribosome, this method has interesting potential applications, including designed polymer synthesis.

Recently, Mao and Tian reported a molecular gear based on DNA [93]. The gears are composed of four DNA single strands: one central circular strand (C) and three peripheral linear strands (Pi, $i = 1,2,3$). By adding the complementary ssDNA strands of the teeth strands, gears could continuously roll against each other (Fig. 12b). Another variation of the movement termed "DNA walker" was recently demonstrated by Shin and Pierce [94]. As described in Fig. 12c, specific DNA strands are used to connect the single-stranded extensions to the labels on the track. The connector strands are equipped with single-stranded toehold sections and can be displaced from the device by their complementary strand via branch migration. This can be repeated several times with the appropriate connector and removal strands to move the walker to arbitrary addresses on the track. Besides the spacial aspects, another report demonstrated that a DNA-based device can control its movement cycle time. A tunable nanomechanical device, called the "nanometronome" has been demonstrated by Ha's team [95]. Their device is made by introducing complementary single-stranded overhangs at the two arms of the DNA four-way junction. The rate of ticking is controlled either by Mg^{2+} or by additional controlling elements, single-stranded deactivator and activator.

4
Double Helix Binding

DNA molecules interact reversibly with a broad range of chemicals that includes water, metal ions and their complexes, small organic molecules, and proteins. Apart from the binding on the backbone along the exterior of the helix, which has been discussed already, there are two locations on the double-helical structure that primarily interact with small molecules: the edges of base pairs in either the major or minor grooves, and the interspace between stacked base pairs. The groove binding (Fig. 1) and intercalation (Fig. 13) are two major modes of interaction between DNA and other compounds. Intercalation results from insertion of a planar aromatic substituent between DNA base pairs, with concomitant unwinding and lengthening of the DNA helix. Groove binding, in contrast, does not perturb the duplex structure to a great extent. Groove-binders are typically crescent-shaped, and fit snugly into the minor groove with little distortion of the DNA structure.

However, such DNA ligands are often toxic or carcinogenic. Nowadays it is well known that several pollutants exhibit carcinogenicity through intercalation into DNA. Some examples are: PAHs or aromatic amines, and some endocrine disruptors, in coal tar, atmospheric pollutants, automobile exhaust, and cigarette smoke. PCDDs, PCDFs, and PCBs have especially emerged as

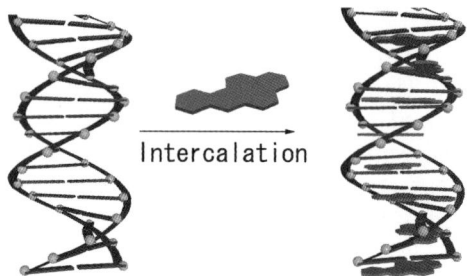

Fig. 13 Schematic representation of DNA intercalation: the small molecules with planar structure intercalate into the interspaces between two adjacent base pairs

potential human carcinogens or endocrine-disrupting chemicals. They exert physiological effects at very low concentrations. It was suggested that DNA be used to remove harmful chemicals before their interaction with human DNA. The DNA-based adsorbents may be have inherent selectivity, as only chemical compounds with special structure would be adsorbed. In contrast, for common adsorbents such as activated carbon and alumina, molecular selectivity is not achieved. However, there is a serious disadvantage in use of DNA for environmental purposes: DNA is highly water-soluble and biochemically unstable. Overcoming these undesirable properties is important for the utilization of DNA as a water cleaning material. In order to manufacture water-insoluble DNA materials, various methods have been reported.

Yamada et al. [96, 97] prepared a dsDNA film with a 3D network by using UV irradiation. A dsDNA solution was applied on glass plates and dried. After treatment with UV light irradiation for more than 1 h, a water-insoluble DNA film was produced. The UV-irradiated DNA films were examined by CD spectra. No change in the CD spectra (max. of 280 nm, min. of 240 nm) was observed, even at a irradiation time of more than 6 h, indicating that UV-irradiated DNA film retains the B-form structure in aqueous solution. To test the biochemical stability, the UV-irradiated DNA films were incubated with *Micrococcal nuclease*. Hydrolyzation results showed that the DNA films have nuclease-resistant characteristics, because of the UV-induced cross-linking between DNA molecules. Although some damages to DNA by UV light have been reported previously, the effect of small changes in the DNA by UV irradiation is not a significant problem for the purpose of carcinogen removal. In subsequent studies, the DNA films were immobilized onto porous glass beads [98] and non-woven cellulose fabric [99] by UV irradiation. These DNA-based materials could effectively accumulate endocrine disruptors and harmful DNA intercalating pollutants, such as dibenzo-*p*-dioxin, dibenzofuran, biphenyl, benzo[a]pyrene and ethidium bromide. The porous beads that had dsDNA immobilized onto the surface by UV irradiation were further used to prepare a DNA-coated glass bead column [98]

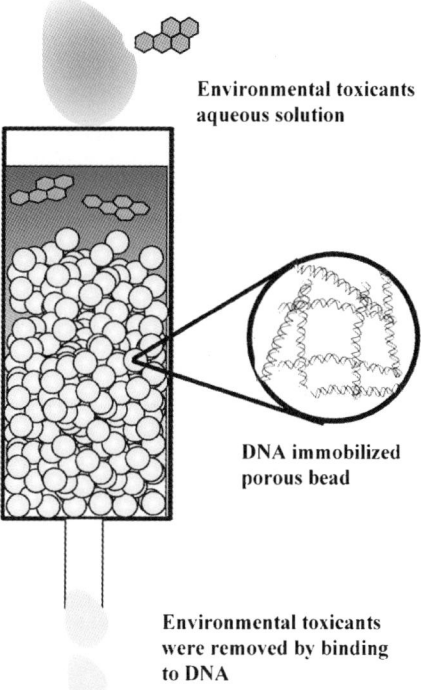

Fig. 14 Schematic illustration of the removal of toxic chemicals by the DNA-immobilized bead column

(Fig. 14). The DNA-immobilized columns effectively accumulated more DNA-intercalating materials than the planar DNA films. The DNA-immobilized columns bound endocrine disruptors with a planar structure, such as dioxins, and benzo[a]pyrene. Bisphenol A and diethylstilbestrol, which lack a planar structure, did not bind to the DNA-coated columns. Table 1 shows the selective adsorptions of the insoluble DNA-based materials.

In another study [100], a dialytic method was proposed for removing and enriching dioxins from polluted water. By combining with a dialytic membrane, the large DNA molecules could be kept inside the dialysis membrane, and then form complexes with small dioxin molecules permeating from the outside. In an experiment, the dioxin derivatives were concentrated in the DNA solution to about 200 times through dialysis of DNA solution in an aqueous mixture of dibenzo-p-dioxin, dibenzofuran, and biphenyl. By mixing with hexane, fluorimetry indicated that the dioxin derivatives intercalated in the DNA molecules leave the DNA solution and transfer to the organic phase, hence, the adsorption capability of the DNA solution was renewed. Although the DNA solution was treated with hexane six times, the removal rates of both dibenzofuran and biphenyl maintained at more than 90% and did not

Table 1 Relative accumulation of endocrine disruptors and harmful compounds by UV-irradiated DNA films (I), DNA-immobilized porous glass beads (II), and DNA-immobilized columns (III)

Chemical compound	Concentration (μM)	Removal rate I (%)	II (%)	III (%)
Dibenzo-*p*-dioxin	0.67	64.7	73.5	87.3
Dibenzofuran	1.1	60.4	69.2	83.0
Biphenyl	1.4	49.8	76.5	97.4
Benzo[a]pyrene	1.4	45.4	66.7	35.7
Bisphenol A	23	0	0	0
Diethystylbestrol	2.1	0	0	0

[a] Each value represents the mean of three separation determinations

appear to decrease. These results suggest that the organic solvent does not damage the DNA double-stranded structure and the adsorption capacity for the dioxins was maintained.

For environmental applications, hybridization of DNA with other polymers is a recommendable approach for creating water-insoluble DNA matrices as sorbent. Several polymers have been reported. By coagulation with bivalent calcium ions, the DNA–alginic acid hybrid matrix could be conveniently insolubilized [101]. The DNA in the hybrid complex was found to be able to adsorb intercalating materials such as ethidium bromide. Following this work, a procedure for immobilizing DNA molecules in a semi-IPN polyacrylamide hydrogel was proposed [102]. Salmon milt DNA was immobilized in the hydrogel beads synthesized by an inverse suspension polymerization of acrylamide in the continuous phase of cyclohexane (Fig. 15c). These DNA hydrogel beads are stable in aqueous medium, more than 82% (w/w) of the DNA can be retained in the hydrogel after soaking in water. Comparing with normal adsorbents such as activated carbon and alumina, this DNA matrix showed a selective adsorption of the dioxin derivatives. Column methods were used to compare DNA hydrogel beads, activated carbon, and alumina. As shown in Table 2, although activated carbon and alumina adsorb more dioxins than the DNA hydrogel beads, they do not exhibit selective adsorption. This property makes the DNA hybrid materials advantageous for removing dioxins from a complex environment containing nutritional compounds. Reusability of the hydrogel beads after washing with organic solvent

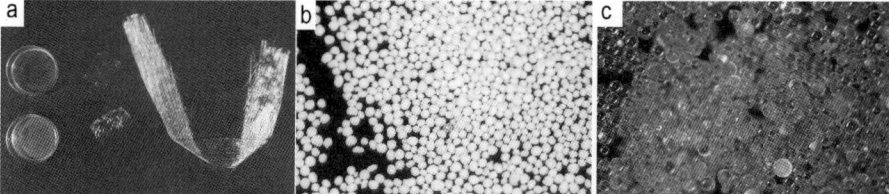

Fig. 15 Optical images of DNA-based materials for environmental purpose. **a** DNA–alginic acid hybrid matrix coagulated by Ca^{2+}, in fiber, film, and gel form. **b** DNA-immobilized porous glass beads prepared by UV-irradiation. **c** DNA–polyacrylamide hydrogel beads synthesized by inverse suspension polymerization

Table 2 Relative accumulation of dioxins and some nutritional compounds by the columns of DNA hydrogel bead (I), activated carbon (II), and alumina (III)

Chemical compound	Concentration	Removal rate [a]		
		I	II	III
	(μM)	(%)	(%)	(%)
Dibenzo-*p*-dioxin	0.65	95.1	98.4	96.5
Dibenzofuran	0.60	95.4	96.2	95.2
Biphenyl	0.60	93.0	95.8	94.3
Vitamin B2	0.75	13.0	93.1	91.0
Vitamin B12	0.75	10.6	95.1	90.8

such as hexane was also confirmed. This is important for the automation and economization of the dioxin removal treatment.

In addition [103, 104], a new type of composite that combines DNA with silica components via a sol–gel method was described. The DNA–silica hybrid material is advantageous with respect to its mechanical and chemical stability in both aqueous and organic solvents. Similar to the previously described hybrids, the specific functions of the DNA molecules were retained and maintained: the DNA–silica hybrid materials adsorb DNA-interactive chemicals from diluted aqueous solution. In another series of reports [105–109], DNA-loaded PSf microspheres were fabricated by means of a liquid–liquid phase separation technique. The release rate of DNA from the microspheres can be controlled by manipulating the microsphere structure. Increasing the polymer concentration causes lower porosity and smaller pores on the outer surface of the microspheres, and leads to a low release rate of DNA from the microspheres. The DNA-loaded PSf microspheres could effectively accumulate harmful DNA-intercalating pollutants and endocrine disruptors, as described in previous reports.

Beside these environmental applications, the intercalation of planar chemicals has been utilized to immobilize DNA. Shimomura et al. succeeded in immobilizing DNA on a Langmuir–Blodgett monolayer using an amphiphilic intercalator, octadecyl acridine orange, at the air–water interface [110]. In a following work, Nakamura and coworkers reported a protocol on the immobilization of DNA molecules onto monolayers containing anthryl groups [111]. Maeda et al. used vinyl derivatives of DNA intercalator to immobilize dsDNA in hydrogels [112, 113]. Recently, Kelley et al. [114] described a DNA-surface immobilization achieved through an ethidium derivative linker on the film surface. The highly conjugated linkage resulted in more efficient electron transfer relative to systems utilizing conventional insulating tethers. The DNA linkers applied in these studies have a similar character, a linear molecule terminated by intercalator and a functional group with high affinity to the solid.

DNA is polymorphic and exists in a variety of distinct conformations. Duplex DNA can adopt a variety of sequence-dependent secondary structures that range from the canonical right-handed B-form through the left-handed Z conformation. Consequently, using DNA as a template for synthesis of supramolecular polymers may give multifarious helical materials. A descriptive example of this approach was published by Armitage et al. [115]. They used an assemblage of symmetrical cationic cyanine dyes bound in the minor groove to produce a helical supramolecular structure.

DNA is chiral by virtue of the asymmetric centers in the ribose units and as a result of the twist of the helix axis. Because of its inherent chirality, DNA is an attractive scaffold for enantioselection, which is of crucial importance in various fields such as drug and food analysis, biochemistry, or clinical pharmacology. Chaires et al. [116] reported a dramatic experiment demonstrating structural selectivity in DNA binding for the naturally occurring anticancer agent (+)-daunorubicin and its (−)-enantiomer. The binding interactions with right- and left-handed DNA have been studied quantitatively by equilibrium dialysis, fluorescence spectroscopy, and circular dichroism. Results indicated that (+)-daunorubicin binds selectively to right-handed DNA, whereas the enantiomer binds selectively to left-handed DNA. Recently, Feringa et al. [117, 118] introduced a novel DNA-based asymmetric catalysis concept. This catalytic ensemble comprises a copper complex of a non-chiral ligand, which incorporates a metal binding site, a spacer, and a covalently attached intercalator, i.e., 9-amino acridine. As a result, the active Cu(II) center is brought into the proximity of the chiral environment of the DNA double helix, allowing for a transfer of chirality from DNA to the reaction product. With this approach, the copper-catalyzed Diels–Alder reaction achieved very high enantioselectivities in water.

The chirality of DNA was applied to selective separation. A DNA aptamer as a new target-specific chiral selector for HPLC was investigated by Michaud et al. [119, 120]. They showed that a tailor-made chiral stationary phase based on a DNA aptamer with known stereospecific binding for the D-enantiomer of the oligopeptide, arginine–vasopressin, exhibits enantioselectivity between the D- and L-peptides. This DNA-based target-specific aptamer chiral stationary phase provides a powerful tool for the resolution of small (bioactive) molecule enantiomers.

5
Biopolymer

DNA is a native substance widely existing in organisms. Irrespective of its genetic information, DNA possesses biophysical and biochemical properties that have been optimized over billions of years of evolution. These unique properties of DNA offer it excellent prospects for serving as a construction material in bioscience. The primary advantage of DNA for bio-applications is based on the fact that DNA is biocompatible. The molecular structure of DNA in vertebrate species is homogenous [121], unlike other biopolymers such as proteins and sugars, and the non- or low immunogenic properties of DNA may limit both innate and acquired immune responses [122]. The second advantage of using DNA as a biomaterial is its perfect binding properties, which have been described in previous sections. Not only other hybrid construction materials but also some pharmacological molecules can be incorporated into DNA via electrostatic inaction, groove binding, and intercalation. Additionally, DNA is degradable by various deoxyribonucleases present in human cells and the digestive system. This feature provides a possibility for using DNA as a protective biomaterial for food or drug delivery applications.

Several attempts have been made to combine DNA with other polymer matrices, cationic lipids, and inorganic supports through a variety of techniques that include physical adsorption, electrostatic binding, and covalent coupling. These materials, which are modified with DNA/oligonucleotides, have been tried for numerous biotechnological applications: improvement of surface compatibility, dental and medical implantology, and cell cultures.

DNA has been used to modify the PSf membrane by blending and immobilizing DNA onto its surface [123, 124]. PSf is one of the most important polymeric materials and is widely used in artificial and medical devices. However, when used as a hemodialysis hollow fiber, the blood compatibility of the PSf membrane is not adequate. The hydrophilicity of the DNA-modified surface increased, but the amount of adsorbed protein did not decrease significantly, which indicates that the DNA-modified membrane might have a better blood compatibility.

Jansen et al. fabricated multilayered DNA coatings on titanium substrates using the electrostatic self-assembly technique, with poly-D-lysine or poly(allylamine hydrochloride) as the cationic counterparts of anionic DNA [125, 126]. In vitro experiments with rat primary dermal fibroblasts indicated that the presence of multilayered DNA coatings do not affect RDF cell viability but did increase proliferation. An in vivo rat model experiment on implants inserted into soft tissue revealed that the presence of a multilayered DNA coating did not induce any adverse effects in terms of inflammation and wound healing. The cyto- and histocompatibility of multilayered DNA coatings demonstrated in this study allows their use and functionalization with appropriate compounds to modulate cell and tissue responses in dental and medical implantology (Fig. 16). The DNA–chitosan complexes have also been investigated for biotechnology applications. The DNA–chitosan bilayer membranes were prepared by applying chitosan solution on UV-irradiated DNA membranes [127]. For the purpose of wound therapeutic application, the bonding strength of the membranes to rabbit peritoneum was tested. These membranes showed an adhesive property to rabbit peritoneum tissue (Fig. 17), and the DNA in the layer retained the double stranded structure. The specific property of DNA in the bilayer membrane can be used as a drug carrier or reservoir. Another work on the DNA–chitosan complexes showed no cytotoxicity for MG-63 osteoblast-like cells and caused only a mild tissue response when implanted subcutaneously in the backs of rats [128].

In other reports, Salmon milt DNA was utilized in oral delivery applications to protect functional materials that are sensitive to the gastric acidity [129–131]. By forming a DNA-based complex gel with gelatin and carrageenan, lactic acid-producing bacteria were protected from extreme acidic conditions. The DNA-based complex gel also showed the desirable conditions for bacterial growth and division in the simulated intestinal juice. Since the electrostatic interaction of the DNA–gelatin complex behaves differently in the simulated intestinal juice and the simulated the gastric condition, this delivery system provided a high protective capability and showed effective results for the survival of the probiotic bacteria.

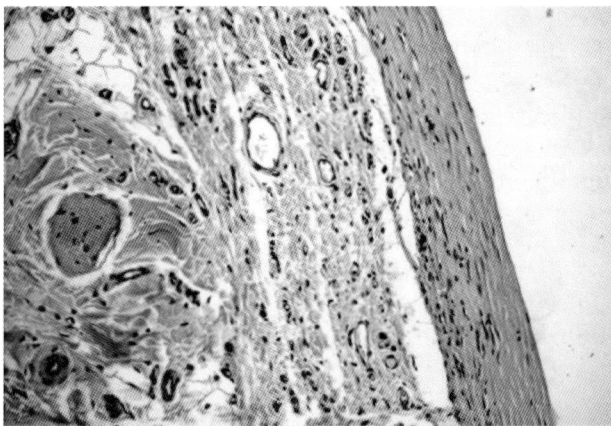

Fig. 16 Light microscopic images of tissue capsule surrounding titanium implants with a DNA and poly-D-lysine complex coating. After 12 weeks of implantation, all implants were surrounded by a relatively mature fibrous capsule without the presence of inflammatory cells at the implant interface. Reprinted with permission from [126]

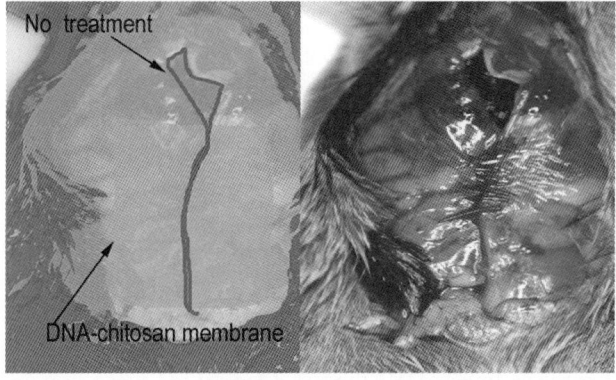

Fig. 17 Optical images of the DNA–chitosan membrane bound onto rabbit peritoneum. The DNA–chitosan complex membrane showed an adhesive property to rabbit peritoneum tissue

Moreover, a noteworthy research tendency is that many recent studies have shown that oligonucleotides possess bioactivity. For example, in a series of three papers, Goukassian et al. have reported on the interdependence of DNA oligonucleotide therapy, skin cancer, and DNA repair capacity [132–134]. The results indicate that treatment of skin cells with thymidine dinucleotide (pTT) increases the DNA repair capacity in vitro and reduces photocarcinogenesis in UV-irradiated hairless mice. Their data further suggest that the age-related decline in DNA repair capacity is substantially reversible, at least in fibroblasts, by treatment with oligonucleotides. Application of such oligonucleotides may enhance DNA repair capacity in human skin, in the absence of actual DNA damage that normally induces the protective response, and thus reduce the carcinogenic risk from solar UV irradiation, particularly in the elderly. In another reports, Sudo et al. provided clear evidence that dietary DNA may play an important role in promoting a shift in Th1–Th2 balance toward Th1-dominant immunity [135–137]. These findings on the bioactivity of DNA might lead to biomaterials derived from DNA with new uses and new challenges to explore, and possibly open new horizons of DNA-based biomaterials.

6
Perspective

In this review, we have presented an overview of the current state of DNA-based functional materials. It is noticeable that many biology techniques contributed to DNA-based materials. In the research on DNA nanostructure, the DNA resource profits from the ability to synthesize virtually any DNA sequence by automated methods and to amplify any DNA sequence in microscopic to macroscopic quantities using PCR. The biology techniques also offer researchers a variety of DNA sample. For example, in Mao's report [72], micrometer-long DNA molecules were synthesized by rolling-circle DNA synthesis. This long DNA template is difficult to obtain from either automated DNA synthesis or nature. Furthermore, discoveries in the biosciences always bring us new inspirations for DNA applications. Liao and Seeman [92] present a DNA device that can program the synthesis of linear polymers through positional alignment of reactants. In this ribosome-like DNA device, there is no complementary relationship between the signal sequence and the products. Like these instances, it is foreseeable that functional materials-based DNA will obtain most power from biosciences and biotechnology. On the other hand, DNA-based adsorbents were considered to have various potential applications in the environmental and health fields. Our results have shown that DNA can be used as an environmental material to selectively remove some toxic pollutants from aqueous solution. However, the possibility of removing pollutants in air by using DNA-based materials is still a challenge to explore. Similar to the behavior in aqueous solution, as discussed previously, it is recognized that DNA has inherent properties for this purpose and it is possible to create many further applications. For example, DNA-containing cigarette filters are hopeful for removing a lot of toxic chemicals such as planar PAHs from the smoke, and a DNA-modified filter can be expected to remove some carcinogenic pollutants in car exhaust gases. Finally, in summary, the use of DNA as a material does not have a long history, and it is still a great challenge to develop DNA-based materials for practical use.

Acknowledgements We thank Professor Olaf Karthaus (Chitose Institute of Science and Technology) for his thoughtful review of the manuscript.

References

1. Marko JF, Cocco S (2003) Phys World 16:37
2. Yamada M, Yokota M, Kaya M, Satoh S, Jonganurakkun B, Nomizu M, Nishi N (2005) Polymer 46:10102
3. Lebrun A, Lavery R (1996) Nucleic Acids Res 24:2260
4. Allemand J, Bensimon D, Jullien L, Bensimon A, Croquette V (1997) Biophys J 73:2064
5. Ferree S, Blanch HW (2003) Biophys J 85:2539
6. Kaji N, Ueda M, Baba Y (2002) Biophys J 82:335
7. Ladoux B, Doyle PS (2000) Europhys Lett 52:511
8. Braun E, Eichen Y, Sivan U, Ben-Yoseph G (1998) Nature 391:775
9. Keren K, Krueger M, Gilad R, Ben-Yoseph G, Sivan U, Braun E (2002) Science 297:72
10. Richter J, Seidel R, Kirsch R, Mertig M, Pompe W, Plaschke J, Schackert HK (2000) Adv Mater 12:507
11. Ford WE, Harnack O, Yasuda A, Wessels JM (2001) Adv Mater 13:1793
12. Mertig M, Colombi Ciacchi L, Seidel R, Pompe W, De Vita A (2002) Nano Lett 2:841
13. Monson CF, Woolley AT (2003) Nano Lett 3:359
14. Gu Q, Cheng CD, Haynie DT (2005) Nanotechnol 16:1358
15. Nakao H, Shiigi H, Yamamoto Y, Tokonami S, Nagaoka T, Sugiyama S, Ohtani T (2003) Nano Lett 3:1391
16. Rainey JK, Goh MC (2002) Protein Sci 11:2748
17. Kitamura H, Iwamoto C, Sakairi N, Tokura S, Nishi N (1997) Int J Biol Macromol 20:241
18. Kaya M, Toyama Y, Kubota K, Nodasaka Y, Ochiai M, Nomizu M, Nishi N (2005) Int J Biol Macromol 35:39
19. Clarkson BH, McCurdy SP, Gaz D, Hand AR (1993) Archs Oral Biol 38:737
20. Mrevlishvili GM, Svintradze DV (2005) Int J Biol Macromol 35:243
21. Mrevlishvili GM, Svintradze DV (2005) Int J Biol Macromol 36:324
22. Ijiro K, Okahata Y (1992) J Chem Soc Chem Commun, p 1339
23. Okahata Y, Ijiro K, Matsuzaki Y (1993) Langmuir 9:19
24. Tanaka K, Okahata Y (1996) J Am Chem Soc 118:10679
25. Okahata Y, Kobayashi T, Tanaka K (1996) Langmuir 12:1326
26. Okahata Y, Tanaka K (1996) Thin Solid Films 285:6
27. Fukushima T, Hayakawa T, Inoue Y, Miyazaki K, Okahata Y (2004) Biomaterials 25:5491
28. Okahata Y, Kobayashi T, Tanaka K, Shimomura M (1998) J Am Chem Soc 120:6165
29. Nakayama H, Ohno H, Okahata Y (2001) Chem Commun, p 2300
30. Kawabe Y, Wang L, Horinouchi S, Ogata N (2000) Adv Mater 12:1281
31. Wang L, Yoshida J, Ogata N, Sasaki S, Kajiyama T (2001) Chem Mater 13:1273
32. Kawabe Y, Wang L, Nakamura T, Ogata N (2002) Appl Phys Lett 81:1372
33. Liang H, Angelini TE, Ho J, Braun PV, Wong GC (2003) J Am Chem Soc 125:11786
34. Radler JO, Koltover I, Salditt T, Safinya CR (1997) Science 275:810
35. Koltover I, Salditt T, Safinya CR (1999) Biophys J 77:915

36. Koltover I, Wagner K, Safinya CR (2000) Proc Natl Acad Sci USA 97:14046
37. Wiethoff CM, Gill ML, Koe GS, Koe JG, Middaugh CR (2002) J Biol Chem 277:44980
38. Natali F, Castellano C, Pozzi D, Castellano AC (2005) Biophys J 88:1081
39. McManus JJ, Radler JO, Dawson KA (2004) J Am Chem Soc 126:15966
40. Yakovchuk P, Protozanova E, Frank-Kamenetskii MD (2006) Nucleic Acids Res 34:564
41. Zhang SW, Fu TJ, Seeman NC (1993) Biochemistry 32:8062
42. Fu TJ, Seeman NC (1993) Biochemistry 32:3211
43. Seeman NC (2003) Nature 421:427
44. Seeman NC (2003) Chem Biol 10:1151
45. Mao C, Sun W, Seeman NC (1999) J Am Chem Soc 121:5437
46. Liu Y, Yan H (2005) Small 1:327
47. Malo J, Mitchell JC, Venien-Bryan C, Harris JR, Wille H, Sherratt DJ, Turberfield AJ (2005) Angew Chem Int Ed 44:3057
48. Seeman NC, Zhang Y, Chen J (1994) J Vac Sci Technol A 12:1895
49. Winfree E, Liu F, Wenzler LA, Seeman NC (1998) Nature 394:539
50. Shen ZY, Yan H, Wang T, Seeman NC (2004) J Am Chem Soc 126:1666
51. Zhang X, Yan H, Shen Z, Seeman NC (2002) J Am Chem Soc 124:12940
52. LaBean TH, Yan H, Kopatsch J, Liu F, Winfree E, Reif JH, Seeman NC (2000) J Am Chem Soc 122:1848
53. Reishus D, Shaw B, Brun Y, Chelyapov N, Adleman L (2005) J Am Chem Soc 127:17590
54. Park SH, Barish R, Li HY, Reif JH, Finkelstein G, Yan H, LaBean TH (2005) Nano Lett 5:693
55. Ke Y, Liu Y, Zhang J, Yan H (2006) J Am Chem Soc 128:4414
56. Mathieu F, Liao SP, Kopatscht J, Wang T, Mao CD, Seeman NC (2005) Nano Lett 5:661
57. Ding BQ, Sha RJ, Seeman NC (2004) J Am Chem Soc 126:10230
58. Yang X, Wenzler LA, Qi J, Li X, Seeman NC (1998) J Am Chem Soc 120:9779
59. He Y, Mao CD (2006) Chem Commun, p 968
60. Chelyapov N, Brun Y, Gopalkrishnan M, Reishus D, Shaw B, Adleman L (2004) J Am Chem Soc 126:13924
61. He Y, Tian Y, Chen Y, Deng ZX, Ribbe AE, Mao CD (2005) Angew Chem Int Ed 44:6694
62. Park SH, Pistol C, Ahn SJ, Reif JH, Lebeck AR, Dwyer C, LaBean TH (2006) Angew Chem Int Ed 45:735
63. Liu Y, Ke YG, Yan H (2005) J Am Chem Soc 127:17140
64. Park SH, Yan H, Reif JH, LaBean TH, Finkelstein G (2004) Nanotechnol 15:S525
65. Shih WM, Quispe JD, Joyce GF (2004) Nature 427:618
66. Rothemund PW (2006) Nature 440:297
67. Samori B, Zuccheri G (2005) Angew Chem Int Ed 44:1166
68. Niemeyer CM, Simon U (2005) Eur J Inorg Chem, p 3641
69. Storhoff JJ, Elghanian R, Mucic RC, Mirkin CA, Letsinger RL (1998) J Am Chem Soc 120:1959
70. Alivisatos AP, Johnsson KP, Peng X, Wilson TE, Loweth CJ, Bruchez MP Jr, Schultz PG (1996) Nature 382:609
71. Loweth CJ, Caldwell WB, Peng X, Alivisatos AP, Schultz PG (1999) Angew Chem Int Ed 38:1808
72. Deng Z, Tian Y, Lee SH, Ribbe AE, Mao C (2005) Angew Chem Int Ed Engl 44:3582
73. Liu D, Daubendiek SL, Zillman MA, Ryan K, Kool ET (1996) J Am Chem Soc 118:1587

74. Koplin E, Niemeyer CM, Simon U (2006) J Mater Chem 16:1338
75. Chaize B, Nguyen M, Ruysschaert T, le Berre V, Trevisiol E, Caminade AM, Majoral JP, Pratviel G, Meunier B, Winterhalter M, Fournier D (2006) Bioconjugate Chem 17:245
76. Mirkin CA, Letsinger RL, Mucic RC, Storhoff JJ (1996) Nature 382:607
77. Rogers PH, Michel E, Bauer CA, Vanderet S, Hansen D, Roberts BK, Calvez A, Crews JB, Lau KO, Wood A, Pine DJ, Schwartz PV (2005) Langmuir 21:5562
78. Milam VT, Hiddessen AL, Crocker JC, Graves DJ, Hammer DA (2003) Langmuir 19:10317
79. Kim JA, Biancaniello PL, Crocker JC (2006) Langmuir 22:1991
80. Li M, Mann S (2004) J Mater Chem 14:2260
81. Brody EN, Willis MC, Smith JD, Jayasena S, Zichi D, Gold L (1999) Mol Diagn 4:381
82. Liu Y, Lin CX, Li HY, Yan H (2005) Angew Chem Int Ed 44:4333
83. Li HY, Park SH, Reif JH, LaBean TH, Yan H (2004) J Am Chem Soc 126:418
84. Yan H, Park SH, Finkelstein G, Reif JH, LaBean TH (2003) Science 301:1882
85. Sharma J, Chhabra R, Liu Y, Ke YG, Yan H (2006) Angew Chem Int Ed 45:730
86. Mao C, Sun W, Shen Z, Seeman NC (1999) Nature 397:144
87. Liu DS, Bruckbauer A, Abell C, Balasubramanian S, Kang DJ, Klenerman D, Zhou DJ (2006) J Am Chem Soc 128:2067
88. Liu DS, Balasubramanian S (2003) Angew Chem Int Ed 42:5734
89. Chen Y, Lee SH, Mao C (2004) Angew Chem Int Ed Engl 43:5335
90. Viasnoff V, Meller A, Isambert H (2006) Nano Lett 6:101
91. Yan H, Zhang X, Shen Z, Seeman NC (2002) Nature 415:62
92. Liao SP, Seeman NC (2004) Science 306:2072
93. Tian Y, Mao C (2004) J Am Chem Soc 126:11410
94. Shin JS, Pierce NA (2004) J Am Chem Soc 126:10834
95. Buranachai C, McKinney SA, Ha T (2006) Nano Lett 6:496
96. Yamada M, Kato K, Nomizu M, Sakairi N, Ohkawa K, Yamamoto H, Nishi N (2002) Chem Eur J 8:1407
97. Yamada M, Kato K, Nomizu M, Haruki M, Ohkawa K, Yamamoto H, Nishi N (2002) Bull Chem Soc Jpn 75:1627
98. Yamada M, Kato K, Nomizu M, Ohkawa K, Yamamoto H, Nishi N (2002) Environ Sci Technol 36:949
99. Yamada M, Kato K, Shindo K, Nomizu M, Haruki M, Sakairi N, Ohkawa K, Yamamoto H, Nishi N (2001) Biomaterials 22:3121
100. Liu XD, Murayama Y, Yamada M, Nomizu M, Matsunaga M, Nishi N (2003) Int J Biol Macromol 32:121
101. Iwata K, Sawadaishi T, Nishimura S, Tokura S, Nishi N (1996) Int J Biol Macromol 18:149
102. Liu XD, Murayama Y, Nomizu M, Matsunaga M, Nishi N (2005) Int J Biol Macromol 35:193
103. Satoh S, Yamada M, Nomizu M, Nishi N (2003) Polymer J 35:872
104. Satoh S, Fugetsu B, Nomizu M, Nishi N (2005) Polymer J 37:94
105. Zhao CS, Liu XD, Nomizu M, Nishi N (2004) J Colloid Interface Sci 275:470
106. Zhao CS, Liu XD, Nomizu M, Nishi N (2004) Sep Sci Technol 39:3043
107. Zhao CS, Liu XD, Nomizu M, Nishi N (2004) J Microencapsulation 21:283
108. Zhao CS, Shi L, Xie XY, Sun SD, Liu XD, Nomizu M, Nishi N (2005) Adsorpt Sci Technol 23:387
109. Zhao CS, Yang KG, Liu XD, Nomizu M, Nishi N (2004) Desalination 170:263
110. Nakamura F, Ijiro K, Shimomura M (1998) Thin Solid Films 329:603

111. Nakamura F, Mitsui K, Hara M, Kraemer S, Mittler S, Knoll W (2003) Langmuir 19:5823
112. Umeno D, Kawasaki M, Maeda M (1998) Bioconjugate Chem 9:719
113. Umeno D, Kano T, Maeda M (1998) Anal Chim Acta 365:101
114. Taft BJ, Lapierre-Devlin MA, Kelley SO (2006) Chem Commun, p 962
115. Hannah KC, Armitage BA (2004) Acc Chem Res 37:845
116. Qu X, Trent JO, Fokt I, Priebe W, Chaires JB (2000) Proc Natl Acad Sci USA 97:12032
117. Roelfes G, Feringa BL (2005) Angew Chem Int Ed 44:3230
118. Roelfes G, Boersma AJ, Feringa BL (2006) Chem Commun, p 635
119. Michaud M, Jourdan E, Villet A, Ravel A, Grosset C, Peyrin E (2003) J Am Chem Soc 125:8672
120. Michaud M, Jourdan E, Ravelet C, Villet A, Ravel A, Grosset C, Peyrin E (2004) Anal Chem 76:1015
121. Krieg AM (2000) Vaccine 19:18
122. Krieg AM, Yi AK, Matson S, Waldschmidt TJ, Bishop GA, Teasdale R, Koretzky GA, Klinman DM (1995) Nature 374:546
123. Zhao CS, Liu XD, Nomizu M, Nishi N (2003) Biomaterials 24:3747
124. Zhao CS, Liu XD, Rikimaru S, Nomizu M, Nishi N (2003) J Membrane Sci 214:179
125. van den Beucken J, Vos MRJ, Thune PC, Hayakawa T, Fukushima T, Okahata Y, Walboomers XF, Sommerdijk N, Nolte RJM, Jansen JA (2006) Biomaterials 27:691
126. van den Beucken J, Walboomers XF, Vos MRJ, Sommerdijk N, Nolte RJM, Jansen JA (2006) J Biomed Mater Res Part A 77A:202
127. Rikimaru S, Wakabayashi Y, Nomizu M, Nishi N (2003) Polymer J 35:255
128. Fukushima T, Hayakawa T, Okamura K, Takeda S, Inoue Y, Miyazaki K, Okahata Y (2006) J Biomed Mater Res Part B 76B:121
129. Jonganurakkun B, Nodasaka Y, Sakairi N, Nishi N (2006) Macromol Biosci 6:99
130. Jonganurakkun B, Liu XD, Nodasaka Y, Nomizu M, Nishi N (2003) J Biomater Sci Polym Ed 14:1269
131. Sundaram S, Viriyayuthakorn S, Roth CM (2005) Biomacromolecules 6:2961
132. Goukassian DA, Helms E, van Steeg H, van Oostrom C, Bhawan J, Gilchrest BA (2004) Proc Natl Acad Sci USA 101:3933
133. Goukassian DA, Bagheri S, El-Keeb L, Eller MS, Gilchrest BA (2002) FASEB J, p 16
134. Goukassian DA, Helms E, Arad S, Kobayashi N, Mori T, Gilchrest BA (2002) J Invest Dermatol 119:330
135. Sudo N, Aiba Y, Oyama N, Yu XN, Matsunaga M, Koga Y, Kubo C (2004) Int Arch Allergy Immunol 135:132
136. Sudo N, Aiba A, Takaki K, Tanaka K, Yu XN, Oyama N, Koga Y, Kubo C (2000) Clin Exp Allergy 30:979
137. Matsunaga M, Ohtaki H, Takaki A, Iwai Y, Yin L, Mizuguchi H, Miyake T, Usumi K, Shioda S (2003) Neurosci Res 47:269

Editor: Shiro Kobayashi

Engineering Blood-Contact Biomaterials by "H-Bond Grafting" Surface Modification

Dong-An Wang

Division of Bioengineering, School of Chemical & Biomedical Engineering, Nanyang Technological University, 70, Nanyang Drive, N1.3-B2-13, 637457 Singapore, Singapore
dawang@ntu.edu.sg

This literature is contributed in memory of my beloved advisor, Professor Lin-Xian Feng of Department of Polymer Science, Zhejiang University, China

1	Polymeric Biomaterials for Cardiovascular Medical Applications	182
1.1	Cardiovascular Biomedical Materials	182
1.1.1	Biomaterials and Cardiovascular Applications	182
1.1.2	Categories of Cardiovascular-Functional Polymers	183
1.1.3	Technical Parameters of Cardiovascular Biomaterials	184
1.2	Surface Optimization and Protein Adsorption	185
1.2.1	Protein Adsorption	185
1.2.2	Biological Recognition of Cell Adhesion	186
1.2.3	Biomaterial Surface Construction for Optimal Protein Binding	187
1.3	Surface Thrombogenesis and Thromboresistance	189
1.3.1	Blood–Implant Interaction	189
1.3.2	Anti-Thrombogenic Coating	190
1.4	Surface-Engineered Endothelialization	192
1.4.1	Functions of Endothelium	192
1.4.2	Engineered Endothelialization for Artificial Implants	193
1.5	Surface Immobilization of Proteinic Affinity Ligands	195
1.5.1	Surface Bioconjugation	195
1.5.2	Spacer Arm	199
2	"H-Bond Grafting" Surface Modification for Blood-Contact Polyurethane Biomaterials	200
2.1	"H-Bond Grafting" Surface Modification for Polyurethane Biomaterials	200
2.1.1	Model Materials and Surface Modifying Additives [SMA]	200
2.1.2	SMA "H-Bond Grafting" for Stable Blend and Surface Modification	203
2.1.3	SMA Migration and Surface Enrichment	205
2.1.4	SMA Surface Conformation and Optimization	207
2.2	Bio-Functionality of Engineered Polyurethane Surfaces by "H-Bond Grafting"	210
2.2.1	Coating Applications and In Vivo Tissue Compatibility	210
2.2.2	Selective Protein Binding	211

2.2.3 Thromboresistance 215
2.2.4 Endothelialization 219

3   Conclusions 221

References 222

Abstract This review consists of two major parts: recollection of advances in therapeutic cardiovascular biomaterials and a summary of "H-bond grafting" methodology for biocompatible/biofunctional surface modification of blood-contact polyurethanes. The development of a H-bond grafting model as depicted in the second part is initiated with originality that is based on an understanding and rendering of advantages extracted from the comprehensive investigations reviewed in the first part. The H-bond grafting strategy is invented via mimicking the buildup of hydrogen bond-based physical crosslinking points in elastomeric polyurethanes, by which the accordingly designed surface-modifying additives are anchored to the virtual interface with the favor of a noncovalent mechanism and the talent of microenvironmental optimization between biomaterials and the biological counterparts. This review assembles a series of self-contained topics covering aspects from prototype setup through various blood-contacting assessments. As a platform of delivery, superior efficacies have been achieved from the H-bond grafting-modified polyurethane surfaces typically on albumin-selective binding, biocompatibility, and engineered endothelialization.

Keywords Biomaterials · Blood compatibility · Surface modification · Polyurethane · Hydrogen bond · Endothelialization · Protein adsorption

Abbreviations
A   adenosine
A(Ala)   alanine
AA   amino acid
ADP   adenosine diphosphate
AT-III   anti-thrombine III
ATR-FT-IR   attenuated total reflection Fourier transform infrared spectroscopy
BSA   bovine serum albumin
C   cytidine
C18   stearyl, stearic group
C(Cys)   cysteine
CDI N,N'-carbonyldiimidazole
Ciba   triazine dye Cibacron Blue F3G-A
CNBr   cyanogens bromide
D(Asp)   aspartic acid
DCC   dicyclohexyl carbodiimide
E(Glu)   glutamic acid
ECM   extracellular matrix
EC   endothelial cell
EDC   1-ethyl-3-(3-dimethylaminopropyl) carbodiimide
EDRF   endothelial-derived relaxing factor
ePTFE   expanded poly(tetraflouro ethylene)
F(Phe)   phenylalanine
Fg   fibrinogen

FGF	fibroblast growth factor
FMP	2-fluoro-1-methylpyridiniumtoluene-4-sulfonate
Fn	fibronectin
G	guanosine
G(Gly)	glycine
H(His)	histidine
H-bond	hydrogen bond
HGB	hemoglobin
H-NMR	proton nuclear magnetic resonance spectroscopy
HSA	human serum albumin
HUVEC	human umbilical vein endothelial cell
I(Ile)	isoleucine
Ig	immunoglobulins
IR	infrared spectroscopy
K(Lys)	lysine
L(Leu)	leucine
Ln	laminin
M(Met)	methionine
MDI	4,4′-methylene diphenyl diisocyanate
Mϕ	macrophage
MPEO	"ABCBA"-type compound: "A" for functional endgroup, "B" for PEG spacer, and "C" for MDI coupling group
MPEO-OH	MPEO template without functional endgroups at the end of PEG spacers
MSPEO	"ABCBA"-type compound: "A" for stearic endgroup, "B" for PEG spacer, "C" for MDI coupling group
MW	molecular weight
N(Asn)	asparaginate
NHS	N-hydroxy succinimide
OM	optical microscope
P(Pro)	proline
PC	polycarbonate
PCL	polycaprolactone
PE	polyethylene
PEO	poly(ethylene oxide)
PEG	poly(ethylene glycol)
PEsU	polyester urethane
PET	poly(ethylene terephthalate)
PEU	polyether urethane
PG(E_1)	prostaglandin (E_1)
PLGA	poly (lactic/glycolic acid)
PRT	plasma recalcification time
PTMG	poly(tetramethylene glycol)
PTT	plasma thromboplastin catalyzed clotting time
PU	polyurethane
PVC	poly (vinyl chloride)
Q(Gln)	glutamine
R(Arg)	arginine
RBC	red blood cells
S(Ser)	serine
SEM	scanning electronic microscope

SK	streptokinase
SMA	surface modifying additive
SMC	smooth muscle cell
SPEO	mono-stearic PEG
T	thymidine
T(Thr)	threonine
TFPI	tissue factor pathway inhibitor
Tg	(polymer) glass transition temperature
Tm	(polymer) melting point temperature
TM	thrombomodulin
t-PA	tissue plasminogen activator
TT	thrombin-catalyzed clotting time
TXA_2	thromboxane A_2
UK	urokinase
u-PA	urokinase-type plasminogen activator
U	uracil
V(Val)	valine
Vn	vitronecti
W(Trp)	tryptophane
WBC	white blood cells
X	xanthosine
XPS	X-ray photoelectron spectroscopy
Y(Tyr)	tyrosine

1
Polymeric Biomaterials for Cardiovascular Medical Applications

1.1
Cardiovascular Biomedical Materials

1.1.1
Biomaterials and Cardiovascular Applications

The concept of biomaterials covers all natural and synthetic non-medicinal materials or blends that, as a whole or partially, fulfill the purposes of physically reconstituting, improving, substituting, or reconstructing defective human tissues or organs [1]. Biomaterials include metal, ceramic, synthetic organic materials, and natural macromolecular materials, of which functional polymers are one of the largest families. The design and fabrication of biomaterials is aimed at a close integration with host biological substances, including proteins, cells, tissues, organs, and even organ systems, functioning for the "engineered therapy" [2]. Therefore, besides the functionalities, actual applications of biomaterials firstly rely on their biocompatibility. Biocompatibility of biomaterials is defined from two aspects: tissue-compatibility and blood-compatibility (or haemo-compatibility). Ideally, tissue-compatibility of biomaterials requires systematic compatibility when contacting the body components, that is, biomaterials must be non-toxic, non-carcinogenic, non-antigenic (for endogenous inflammation), and non-oncogenic (for gene mutant) [3–6]. Potential exogenous infections may be induced by comprehensive factors from surgical and post-surgical operations, as well as some endogenous focus like formation of thrombus that not only provides a breeding ground for disease vectors, but also destroys materials' haemo-compatibility. For blood-contacting biomedical materials, including cardiovascular biomaterials, haemo-compatibility, which pursues the maintenance of blood rheological properties and biochemical compositions by thromboresistance, is a fundamental requirement. Currently cardiovascular biomaterials are employed for various applications including (1) devices for in vitro whole blood containing and transfer; (2) intravascular devices for interventional therapy; and (3) devices for permanent transplantation as soft tissue substitutes. Different applications are endowed with different criteria for haemo-compatibility. For example, materials of heart valves for long-term implantation should sustain their biocompatibility for longer periods than that of instant intravascular guiding catheters [7–10].

1.1.2
Categories of Cardiovascular-Functional Polymers

Cardiovascular-functional polymers cover almost all categories of synthetic polymers and large numbers of biopolymers. They are used to build the device bulk, act as surface-modifying additives [SMA], and also formulate tissue adhesives [11–17].

Among the bulk materials, non-degradable synthetics are usually selected as the primary candidates for fabricating in vitro or interventional devices, such as catheters, sutures, tubing, blood bags, housing materials, wound dressings, angioplasty balloons, etc; and also employed for constructing permanent substitutes, such as an artificial heart, heart bladders, heart valves, etc. Permanent bulk materials include glass-state polyacetals, polyamides (and their elastomers), polycarbonates [PC], polyesters, polyethers (epoxies and their elastomers), polyimides, poly(methylpentene), polyolefins (and their elastomers, and high crystallinity films), polyurethane [PU] elastomers, poly (vinyl chloride) [PVC], ultra high molecular weight polyethylene [PE], acrylics (hydrogels), silicones and fluorocarbons. On the basis of the classic polymers listed above, the recent development of bioresorbable polymers has provided hope of achieving non-secondary-surgical suture removal, controlled drug-releasing substitute scaffolds, and in situ tissue-engineered implants. Controllable degradation or bioresponsive degradation is the key to fulfilling these purposes. Traditional candidates for bioresorbable materials are poly (amino acids), polycaprolactones [PCL], and poly (lactic/glycolic acid) [PLGA] copolymers. Biodegradable polyurethanes and polyphosphoesters are also being examined as to their potential uses.

Biopolymers utilized as bulk materials for medical devices have the advantages of specific bio-functionality and general biocompatibility, but on the other hand, the drawbacks of difficult fabrication and the risk of stimulating immuno-reactions have also limited their application. Two categories of biological materials are developed and employed: native organ derivatives and bio-derived macromolecules. Widely used native organ derivatives are blood vessels (arteries and veins, bovine), pericardium (bovine), umbilical cord (vein, human), and heart valve (porcine) etc; bioderived macromolecules include albumin (crosslinked), cellulose (acetates or cuprammonium), chitosan, collagen, elastin, gelatin (crosslinked), phospholipids, and silk etc.

In contrast with bulk materials that take responsibility for product mechanics and metabolism, the superior blood-contacting functions of cardiovascular biomaterials are conveniently fulfilled by surface modification. The strategy of surface modification is to introduce either passive (inert) or bioactive SMA onto the blood–material interface. By surface optimization, the material is granted with variable functions including thromboresistance, infection resistance, selective protein binding, promotional cell adhesion, and even simply lubricity. Surface-modifying approaches include chemical decoration (via covalent conjugation) and physical coating (blends). Popular passive factors rendered via SMA are albumin, alkyl oligo-chains, fluorocarbons, and silicones (silica-free products and oils) etc; bioactive functional groups immobilized through SMA are usually anticoagulants (like heparin and hirudin), antimicrobials (like chitosan), cell-adhesion peptides/proteins (like RGD tri-peptide and fibronectin [Fn]), plasma components (polymerized coatings), and thrombolytics etc. Tissue adhesives are also used for cardiovascular applications, such as microsurgery for anastomosing vessels and enhancement of cell adhesion. The involved categories are mainly cyanoacrylates, Fibrin glue and molluscan glue, etc.

1.1.3
Technical Parameters of Cardiovascular Biomaterials

From the material fabrication and performance point-of-view, nine aspects of technical parameters are mainly involved [18–20]:

1. **Strength**. Major arteries, heart valves and angioplasty balloons require superior strength of applied materials (using engineering plastic and synthetic fiber composites); guiding catheters, surgical sutures and intravascular stents require medium strength (using common plastic, elastomers, and native organ derivatives); non-supporting and mild-shear-bearing coatings have limited mechanical requirement (using biomacromolecules and hydrogels).
2. **Incision anastomosis**. Long-term interventional catheters, heart or blood vessel patches and artificial vascular grafts need to anastomose well with

the adjacent native tissues minimizing the secondary pathological variations at the site of incision.
3. **Creep-resistance.** Material creep-resistance evaluates the capacity of materials maintaining their shape and size under constant stresses. Chemical and/or physical crosslinking helps materials achieve high creep-resistance, such as thermosetting plastics (by chemical crosslinking) and multi-phased block copolymers (via hydrogen bond [H-bond] crosslinking points, like in polyurethane elastomers).
4. **Aging-resistance.** Aging-resistance specifically describes the capacity of materials to resist critical stress-induced cracking. The majority of plastics or elastomers are lacking in aging-resistance, except for biomedical-grade poly (ether urethanes) and their derivative products.
5. **Swelling-resistance.** Water-absorbing and swelling directly distorts the materials shape and size and reduces their mechanical performance.
6. **Lubricity.** Outer-wall lubricity is particularly important for intravascular catheters, otherwise the endothelial layer of blood vessels is easily damaged by interventional insertion. Lubricity is usually acquired by coating with hydrogels or biopolymers.
7. **Degradation-resistance.** Unexpected material degradation is usually caused by hydrolysis, oxidization, enzymolysis, lipid invasion or calcification. The talent of degradation-resistance is essential for materials used in permanent implants.
8. **Bioactivity.** Bioactivity of materials is defined as the characteristic of inducing, involving or impacting biochemical/physiological reactions among the adjacent biological units, such as bioactive coatings hindering blood coagulation, negatively charged surface modification diminishing the complement activation, and surface peptide/protein hybridization promoting endothelialization or native soft tissue adhesion.
9. **Bioresponses.** In contrast with "bioactivity" that describes how materials affect biological systems, bioresponses focus on the counter-procedure—how biological environment influences materials and stimulates their responses. Applications of material bioresponses include both the resistance of negative (harmful) responses like inflammation or thrombogenesis and the generation of positive (helpful) reactions like bioresponsive degradation or drug release.

1.2
Surface Optimization and Protein Adsorption

1.2.1
Protein Adsorption

Proteins largely and extensively exist in all kinds of living beings, constructing their bodies and also acting as information carriers as well as recognition probes. The manner of protein binding and patterning on biomaterial surfaces directly impacts their biocompatibility and bio-functionality [21–24]. Interactions between soluble proteins and material surfaces are dominated by thermodynamic criterion, namely, the principle of lowest surface (interface) energy and maximal surface entropy. Compared with most types of polymers, water-soluble proteins are usually more hydrophilic and always have some amphiphilic characteristics. When protein-containing fluid contacts polymer surfaces, proteins will spontaneously attach to the foreign interface bridging and isolating the material from the water phase, by which the surface energy is minimized. Simultaneously, the original uniformly arranged intermolecular H-bonds in the superficial layer (interface) of the water phase (because of a lack of interference by the relatively hydrophobic polymer interface) is disturbed by the inserting amide groups of the proteins, which are efficient H-bond "vectors". Consequently, the randomization of H-bond arrangement leads to an increase of interface entropy. Therefore, in general, the more hydrophobic material surface usually induces a stronger tendency for soluble protein adsorption. Besides thermodynamic factors, a material surface's electrical properties also influences protein-adsorbing behavior. Since protein backbones mostly bear negative changes, despite some positively charged peptide side-chains, protein adsorption is always more prone to occur on positively charged material surfaces.

All soluble proteins have their specific 3D conformations in the water phase, which is engendered by the protein-chain intra-molecular polarity and dispersion, plus intermolecular disulfide bridging, and also relates to the interaction with solvents. Usually spontaneous protein adsoption reduces surface energy, while also inevitably accompanies with conformational variation of the committed proteins. If the change is irreversible, variable extents of protein denaturation will occur [25, 26]. How to avoid denaturation of the binding proteins remains a significant topic. Investigations indicate that, for the binding proteins, reversible adsorption frequently hints at reversible conformation changes, which suggests a strategy of dynamically balancing the counteractions of protein adsorption-desorption. Another important finding for multi-species protein adsorption is the predominance of preoccupation. When blood contacts a material's surface, the largest family of serum proteins, albumin, will always reach the interface first and consequently dominate the following manner of serum protein adsorption.

1.2.2
Biological Recognition of Cell Adhesion

No matter whether at the interface of foreign materials or under in vivo conditions, cell adhesion is always accomplished via interactions between extracellular ligands and receptors on the cell surface. Typical species of cell adhesion receptors are integrins, proteoglycans, and selectins. Accordingly, a large amount of soluble proteinic ligands extensively exist in blood or other bodily fluids, including fibrinogen [Fg], Fn, vitronectin [Vn], thrombospondin, and some of the Willebrand factors, etc; besides, some insoluble ligands, such as collagens, laminin [Ln], or high molecular weight Willebrand factors, only exist in the crosslinked macromolecular extracellular matrix [ECM] [27, 28]. Cells have the outstanding ability to reconstitute their ECM. All the proteinic ligands are synthesized and secreted by cells and are also able to be removed by cellular proteolysis [29–37].

1.2.3
Biomaterial Surface Construction for Optimal Protein Binding

Despite of hybridization with biological ligands, the general strategy for optimizing protein adsorption on biomaterial surfaces relies on chemical or physicochemical modulation of surface hydrophilicity [38, 39]. The common approach is surface immobilization of hydrophilic polymers like polyethylene glycol [PEG] or polysaccharides. Four categories of surface-modification pathways have been developed:

1. **Surface covalent grafting.** This strategy usually combines surface conjugation and in situ polymerization. The methods include regular chemical grafting, photochemical grafting and radio-electrochemical grafting, etc. The advantage is reliable conjugation stability, and the drawback is low grafting efficiency that is caused by the preliminarily introduced hydrophilic SMA chains kinetically excluding subsequent grafting. Since for the performance of surface modification, the contribution of sufficient SMA magnitude is equally crucial to SMA chemical properties, the drawback of low SMA density largely limits its application. Novel techniques like plasma multi-layer grafting or hydrophilic dendrimer/microsphere electrophoresis have been proposed. However, the operational complication and residual toxicity still remain bottlenecks [40–49].
2. **Amphiphilic adsorption.** To improve surface grafting efficiency, the alternative way is pre-adsorption of amphiphilic SMA followed by in situ chemical immobilization. Taking advantage of the surfactant effect that reduces surface energy at the material interface, amphiphilic adsorption thermodynamically overcomes the kinetic hindrance for continuous SMA supply and maximizes SMA magnitude in surface area. The grafting stability is also reinforced by chemical crosslinking. Generally, this aqueous-compatible strategy fulfills the requirement of "in situ fabrication-and-in situ use" and thereby makes it convenient for practical applications [50–54].
3. **Surface self-assembly.** On the basis of an amphiphilic adsorption strategy, instead of chemical immobilization, non-covalent supermolecular systems can be constructed by SMA surface self-assembly. The represen-

tative model is self-assembly LB membranes made of amphiphilic PEG derivatives or copolymers. Other typical models established with various mechanisms include the thioalcohol-gold monolayer system based on chemical affinity, and polyelectrolyte/polyion alternative multilayer system including polylysine (polycation) and algin (a negatively charged calcium-containing hydrogel) microcapsules [55–66].

4. **Matrix bulk blending.** Blends of material matrix and SMA in material bulk are also used for biomaterial surface modification. The matrix-material candidates are usually thermoplastic elastomers like polyurethane and degradable glass-state scaffolds like PLA-PGA. For degradable scaffolds, any surface modification easily loses its function at the earliest degradation of the surface layer; contrary to this, bulk blending is always capable of maintaining SMA present at the interface of degrading cracks. For blends in elastomers, using the elastomeric materials' talent of environment-dependent "annealing" characteristics, SMA locational distribution and orientation are re-organizable by their self-migration and re-assembly. Collaborating with the moveable macromolecular chains of the elastomeric matrix, the spontaneous SMA migration is driven by crystallo- or non-crystallo micro-phase separation (between SMA and the matrix material) and oriented by SMA's amphiphilic surface-occupation tendency, which eventually achieves SMA surface enrichment. During the whole procedure, three points of concern are involved: the modulation of the micro-phase separation; the interrelationship between material mechanics and biocompatibility; and the functional contribution by SMA size and chemical properties [67–75]. Firstly, the stability of SMA participation against phase separation needs to be maintained by SMA-matrix crosslinking, but strictly avoiding the loss of the matrix material's elastomeric characteristics. Thereby physical crosslinking, like H-bond connection/incorporation, is preferred. Micro-phase separation in bulk inevitably leads to micro-domain separation on the surface, by which the number, species, and status of binding proteins are varied, and consequently the material biocompatibility as well as the bio-functionality is influenced. Investigations also indicate that material mechanics and biocompatibility are frequently two counter-factors to be optimized; and simultaneously, the equal contributions of SMA size, physical state, and chemical properties for material biocompatibility are further emphasized. An H-bond grafting strategy for polyurethane elastomer surface modification via SMA-matrix bulk blending and its cardiovascular applications are thoroughly reviewed in Sect. 2 [76–83].

1.3
Surface Thrombogenesis and Thromboresistance

1.3.1
Blood–Implant Interaction

When intervening with the host blood, artificial implants (including the so-called "inert materials") are always recognized as foreign invaders tending to induce variable host reactions, such as serum protein precipitation, platelet aggregation, inflammatory leukocyte infiltration, limited endothelial cell [EC] ingrowth and smooth muscle cell [SMC] immigration, etc. These body-responding events directed by miscellaneous mechanisms are crucial factors for successful transplantation of artificial devices.

Serum Protein Adsorption

Serum protein precipitation occurs almost immediately after blood–implant contact. Rich serum-containing proteins—albumin, fibrinogen, and immunoglobulin families [Ig, especially IgG]—dramatically adsorb to material surfaces and reorganize their distribution based on different biochemical and/or electrical affinities with the foreign surfaces, as described by the "Vroman Effect". Since the surface binding of platelets or other blood cells must be intermediated by the pre-occupied serum proteins, the expressed concentration of proteinic ligands at the blood–material interface actually dominates their counterparts activation. Furthermore, instead of the involvement by whole protein macromolecular chains or even their three-dimensional supermolecular complexes, the function of proteinic ligands is frequently executed by some of their oligo-peptide moieties. For example, the proteinic ligands functioning for cell adhesion and cell spreading direction always demonstrate their functionalities via some highly repeating effective amino acid [AA] sequences. For Fn, the effective AA sequences are RGDS, LDV, and REDV; for vitromectin, the effective sequence is RGDV; for Ln A, they are LRGDN and IKVAV; for Ln B1, they are YIGSR and PDSGR; for Ln B2, it is RNIAEIIKDA; for type I collagen, they are RGDT and DGEA; and for thrombospondin, they are RGD and VTXG. Typically, the sequence of RGD tri-peptide is extensively present in most of the cell-connecting ligands. Investigations reveal that the domains containing RGD sequences of proteinic ligands are specifically in charge of incorporating the G II-b/III-a receptors on the platelet surface. Besides the blood cells, the pre-bond proteins on implant surfaces also significantly influence the activation of serum complements.

Platelet Precipitation

As described above, besides a certain participation of platelet physical adsorption, the interaction between platelet surface receptors and the pre-bond proteinic ligands on the material surface is mainly responsible for the initial platelet precipitation. The platelet-protein complex is eventually formed via the intermediation of the Willebrand factor, a glycoprotein located on the platelet cell membrane. Subsequently, the precipitated platelets experience conformational change and secrete a series of bioactive substances (hormones) including serotonin, cortin, ADP, and thromboxane $A_2[TXA_2]$, which further activates more platelets and catalyzes the production of thrombin, finally resulting in platelet aggregation and thrombogenesis.

Neutrophil and Monocyte Infiltration
Implantation-related acute inflammatory reactions are usually caused by neutrophil adsorption onto the surfaces of synthetic implants, which is mostly intermediated by the strong promoters of C_5a and Leukotrienc B_4 etc. Monocytes in blood circulation are also prone to adsorb to the sites of blood vessel defects or endothelium regeneration. With the stimulation by miscellaneous activating factors in serum, monocytes tend to differentiate into macrophages [Mϕ] and become one of the major participators for host chronic inflammation.

Limited Endothelial Cell Ingrowth and Smooth Muscle Cell Invasion
The continuous and orderly EC patterning generates the perfect integrity of blood vessel endothelium and simultaneously functions to secrete bioactive matter resisting thrombosis, initiating thrombolysis, and preventing SMC invasion, thereby maintaining the normal rheology of blood flow. However, from the border of synthetic implants suturing with native tissues, the EC ingrowth is usually very limited. The maximal ingrowing depth is no more than $1 \sim 2$ cm and the area around incision anastomosis is frequently covered by injured and/or non-fused EC with abnormally varied biochemical characteristics. Consequently, the uncovered portions of the material surface severely induce in situ blood clotting and the SMC underneath the endothelium is stimulated to overgrow leading to thrombogenic SMC invasion and endothelium hyperplasia [84–87].

1.3.2
Anti-Thrombogenic Coating

By summarizing the potential host reactions occurring at the blood–material interface, the employment of improperly treated artificial implants will result in serious damage and even death of the host. An alternative method relies on applying anti-coagulating/thrombolytic drugs, like heparin, to act against thrombogenesis. However, this inevitably carries the risk of inducing fatal bleeding and even the risk of a host systematic immuno-disorder. Therefore, the fundamental solution still depends on the improvement of materials by surface modification. As discussed in Sect. 1.2.3, considering the dilemma of material biocompatibility and mechanical strength, also balancing the factors of operational complication and manufacture cost, a physical coating is extensively acknowledged as the optimal choice for the implant's surface modification and is widely employed.

On the basis of different effective longevities, anti-thrombogenic coatings are designed to adapt to variable blood-contacting durations.

1. **Anti-adhesion coating.** An anti-adhesion coating is used on devices for short-term applications, such as intravascular catheters. The coating functionality is achieved by non-specifically repulsing attachment of all platelets, proteins, and any other thrombogenic matter. This artificial "inertness" is rendered by enhancing the surface hydrophilicity and lubricity. It is not necessary to involve any biological reagents. Investigations indicate that, for short-term interventional therapy, device surfaces treated with photocure PEG hydrogels for improvement of hydrophilicity and lubricity demonstrated superior thromboresistance compared to their heparin-immobilized counterparts without deliberate lubricity or hydrophilicity treatment [88].
2. **Bioactive coating.** For the purpose of long-term in vivo intervention, like the use of intra-cardiovascular stents, a simple anti-adhesion coating lacks persistence. Introduction of additional bioactive factors becomes essential for reaching the goal. Among a number of relevant biological factors, heparin is the most popular candidate. Heparin is a highly charged biomacromolecule with the talent of specifically hindering thrombin-Xa (activated Blood Coagulation Factor X) interaction and consequently interrupting the thrombogenic evolution of fibrinogen. Clinically, heparin-containing coatings have demonstrated long-term advantages of clotting resistance and acute coagulation delay. Other widely used bioactive factors are phosphocholine, albumin, and plasminogen. Phosphocholine is ubiquitous in blood. As an excellent substitute for cell membrane component, a phosphocholine-immobilized coating is capable of resisting the adhesion and activation of serum proteins or platelets. Serum albumin is a well-known "inert" protein. Albumin-rendered coating functions for, both in vitro and in vivo, prohibiting acute fibrinogen attachment, thereby blocking the formation of fibrin and preventing platelet aggregation. The family of plasminogen-type activators mainly includes plasminogen, polylysine, streptokinase [SK] and urokinase [UK, e.x. "u-PA"— urokinase-type plasminogen activator]. The common function of these activators is thrombolysis, namely, dissolution or destruction of already-solidified thrombus. Multi-functional surface coatings combining thrombolytics and anti-coagulants are usually more competitive for generating better performance of thromboresistance [89–92].
3. **Re-endothelialization promoting coating.** When designing and applying permanent (whole life) implants like artificial vascular grafts, artificial hearts, or artificial heart valves, more attention and effort has been paid to

resolving host immunoreactions, where the perfect thromboresistance remains a bottom-line requirement. The most straightforward strategy is to render host EC covering the man-made surface and isolating the synthetics from host immune recognition. For this purpose, tissue-engineered artificial implants are developed by laying bio-functional surface coatings that promote the ingrowth of host EC and generate engineered endothelium with integrity. The major means to direct cell proliferation and migration for re-endothelialization is to immobilize and pattern cell-adhesion ligands onto the modified surface. As described in Sect. 1.3.1, some oligo-peptides composed of certain AA sequences, instead of the full-sized protein, have been physically coated or chemically tethered onto the artificial surfaces associating the cell seeding or host cell growth orientation [93–96].

1.4
Surface-Engineered Endothelialization

1.4.1
Functions of Endothelium

Vascular ECs are border cells between blood and the substrate of the blood vessel wall. The number of ECs throughout the human (adult) body is 10^{11} with an overall surface area of ~ 6000 m^2 and a gross weight of $1000 \sim 2000$ g. The suspending cells are round and cobblestone-like, but when attaching to the inner vascular wall, the native EC is orderly and continuously arranged into a flat and transparent monolayer with cell borders fusing each other. The individually spread cells are usually in the shape of a diamond or polygon. The dimensions are, respectively, $25 \sim 50$ μm in the direction parallel with blood flow, $10 \sim 15$ μm in the direction perpendicular to blood flow, and 3 μm in thickness [97–99].

Most cardiovascular diseases are initially visualized as EC pathological change including functional deterioration and physical detachment, which triggers a series of subsequent reactions, as listed in Sect. 1.3.1, including serum components (especially lipid) infiltration, inflammatory macrophage recruiting, sub-endothelium SMC invasion, and platelet aggregation. The events and status of EC dramatically influence each pathological stage from initiation to recovery or termination. Therefore, the function of endothelium is far more than a physical barrier between blood and adjacent tissues. It contributes in three major ways:

1. **Reducing vascular permeability.** EC modulates mass exchange between blood and adjacent tissues, and prevents the random infiltration of serum components or blood cells, which mostly occurs when EC divides or receives external stimulation by thrombin/histamine involvement or mechanical (shear) stress.

2. **Balancing thromboresistance and haemostatic functions.** Under normal conditions, EC reserves anti-thrombogenic and thrombolytic functions to maintain normal blood rheology. For this purpose, multiple pathways are involved: (i) synthesizing anti-coagulants like anti-thrombine III [AT-III], thrombomodulin [TM], heparin-like polysaccharides, and (external) tissue factor pathway inhibitor [TFPI]; (ii) producing thrombolytic promoters like tissue plasminogen activator [t-PA] and u-PA; (iii) releasing platelet adhesion and aggregation inhibitors like prostaglandin (6-keto-PGE_1) and 13-hydroxy stearic diene acid; and (iv) secreting vascular smooth muscle relaxation factors like prostacyclin and endothelial-derived relaxing factor [EDRF]. On the other hand, the haemostatic function of EC acts as a protecting reaction that maintains vascular integrity and endothelium continuity under abnormal conditions like trauma or pathological injury. The corresponding events include generation of therapeutic proteins and the release of EC-derived factors to encourage the constriction of vascular smooth muscles, both of which promote the recovery of vascular defects.
3. **Resisting vascular wall cell migration and hyperplasia.** Vascular EC can be induced to resist SMC invasion. For this mechanism, besides the functions of EC-produced multiple factors, EC itself is also capable of migration with the cue of migrating factors secreted by platelets or leukocytes and the shear stress by blood flow. EC migration usually accompanies proliferation, which frequently relates to new vascular formation or defective vascular regeneration [100, 101].

The conclusion drawn from the above-discussed features of native endothelium is that—where feasible—construction of host endothelium on the surface of synthetic implants—i.e. engineered endothelialization—may present a promising pathway for fundamentally resolving the biocompatibility and bio-functionality of artificial cardiovascular transplantation. This proposed biocompatibility is, however, not simply thromboresistance (or blood compatibility) plus blood-cell compatibility, but a real bioactive, self-renewable and specifically functional alliance of tissues and synthetics.

1.4.2
Engineered Endothelialization for Artificial Implants

Over the five decades following invention and application in the clinic of the first artificial vascular graft in the 1950s investigators initially endeavored to develop implantable products using native biological derivatives or synthetic materials that intrinsically possess outstanding mechanical and biological properties without further modification. The typical synthetic products are made of poly(ethylene terephthalate) [PET] or expanded poly(tetraflouro ethylene) [ePTFE]. They are useful for inducing vascular endothelialization, but unfortunately prone to causing thrombosis. Since it is still unrealistic for synthetic materials to produce a completely bio-inert product, both the pros and cons reflected from the interactions between material and blood components must be balanced for an optimal outcome. Research maintained the adoption of superior polymers as implant scaffolds, but gradually improved the strategy by pursuing active involvement of biological factors that may lead the host reactions to tend positively for the expected purpose. One of the model approaches is surface immobilization of proteinic ligands. Currently, the ultimate goal is to develop bio-absorbable artificial vascular implants that may take the responsibility of maintaining vascular biomechanics and integrity, simultaneously guiding and participating in vascular regeneration, until they are eventually replaced in whole or in part by new vascular tissues. Accordingly, the ideal fabrication of vascular implants is proposed to be an in vitro integration of tissue-engineered cellular precursors and bio-absorbable substrates that are hybridized with guider proteins for host cell ingrowth and precursor renewal [102–105].

Surface-engineered endothelialization for artificial cardiovascular implants can be achieved either by allogenic EC seeding or inducing host EC reoccupation. These two strategies may work separately or in coordination.

Allogenic Endothelial Cell Seeding

The original idea of biomaterial surface endothelialization via in situ growth of allogenic EC was first established in 1978. Afterwards, the primary technical barriers for in vitro EC culture were gradually conquered, covering the whole procedure of cell culture from cell harvest, seeding to cell development, and also resolves the maintenance of the endothelial phenotype on artificial surfaces, namely, prevention of dedifferentiation or desquamation. Animal experiments indicate a better fluency of blood flow in the endothelialized artificial vascular system compared to its untreated control. Further investigation also reveals the anti-infective contribution by endothelializaion. However, for clinical applications, the greatest disadvantage of allogenic EC transplantation is the low cell density. The seeded cells cannot sufficiently integrate with the implant surface to build up a full and continuous coverage. To overcome this problem, attempts have been made at in vitro expansion of allogenic cells until confluence and then performing implantation, or relying on the employment of microvascular EC that is endowed with superior proliferation capacity. Nevertheless, significant drawbacks still exist in the form of the extra complication of the cell culture process and the requirement of secondary surgery. They both produce extra toleration of patients and also increase the chance of contamination and infection [106–108].

Inducing Host Endothelial Cell Reoccupation

Theoretically, specific EC growth factors are capable of stimulating the over ECM ingrowth of capillary vessels and thereby promoting endothelialization of artificial vessels. Fibroblast growth factors, FGF-1 (or aFGF) and FGF-2 (or bFGF), are mainly applied to induce host EC proliferation, migration and differentiation, and eventually finalize EC reoccupation for in situ vascular formation on man-made devices. The coordination of FGF promoters and EC-produced proteolytics plays an essential role for angiogenesis by modulating vascular ECM and enabling EC development. As artificial scaffolds, combinations of collagen/fibrin and collagen/PET have been designed and developed. Equivalent to biological scaffolds like elastins of native vessels, the artificial implants also take care of physical support and structure maintenance. Provisionally successful trials have been performed in vivo with animal models [109–112].

1.5
Surface Immobilization of Proteinic Affinity Ligands

1.5.1
Surface Bioconjugation

In order to achieve biological hybridization on polymeric implants, bioactive factors must be introduced onto biomaterial surfaces. Since the largest family of bioactive factors is proteinic affinity ligands, protein-targeted bioconjugate techniques have been well developed. Protein immobilization can be performed by physical or chemical means, as well as some newly invented electrochemical (like ionization) and physicochemical (like self-assembly monolayer) methods. Physical approaches are relatively simple in process with the mechanism focusing on arranged adsorption. Variable strategies have been utilized including electrostatic assembly, Van der Waals interaction, H-bonding incorporation, and hydrophilic/hydrophobic affinities. However, the drawbacks are lack of persistence and also vulnerability by variation of conditions. In contrast, chemical conjugation resolves immobilizing stability by covalent bonding.

The covalent surface immobilization of proteins is usually performed with two steps: substrate (surface) activation and protein-substrate conjugation. The targets of activation on the substrate surface are reactive groups like hydroxyl [– OH], amine [– NH], carbonyl [C = O] or carboxyl [COOH]. If these groups are intrinsically absent in the substrate material, a surface pre-treatment must be carried out to generate them by oxidization, radiation, hydrolysis, thiolysis, amination or aminolysis, etc. Sometimes, additional grafting of more specifically functional units, such as (meth)acryloyl [$CH_2 = C - CO$] (for free radical addition) or vinylsulfone [$CH_2 = CH - SO_2$] (for hydrosulphonyl addition), is also required. On most occasions, the activated substrate covalently connects with the proteins via their pendant amines (mostly provided by lysine, arginine, histidine moieties or side-chain N-termini) or carboxyl (mostly provided by aspartic acid moieties or side-chain C-termini); while on some other occasions, the candidate units for protein conjugation can also be pendant hydrosulphonyl [– SH] (of cystine residues) or phenol (of tyrosine residues). Therefore, the bioconjugate technique for protein immobilization is actually a coupling chemistry between these reactive groups of both sides. Water is usually the only good solvent for most of the soluble proteins, and considering this the Step II (protein-substrate coupling) reaction must be performed in an aqueous medium with mild temperature, pH and salinity conditions [113–120].

The reaction Step I (substrate surface activation) and Step II (protein-substrate coupling) can be intermediated by activating-and-leaving reagents or multi-functional coupling reagents, which demonstrate totally different mechanisms. The activating-and-leaving reagent behaves like a catalyst that renders an active leaving group onto the substrate surface reducing the reactive threshold for the following proteinic ligand substitution; while the multi-functional coupling reagent physically tethers the ligand and substrate together by covalently bonding them both, namely, playing the role of permanent bridge. A combination of these two strategies is also widely used. The popular activating-and-leaving reagents are *N*-hydroxy succinimide esters [NHS] (and water soluble sulfo-NHS), dicyclohexyl carbodiimide [DCC] (and water soluble 1-ethyl-3-(3-dimethylaminopropyl) carbodiimide [EDC]), *N,N*′-carbonyldiimidazole [CDI], sulfonyl chlorides, and 2-fluoro-1-methylpyridinium toluene-4-sulfonate [FMP], etc. The relevant model reactions are listed in Table 1A. The multi-functional coupling strategies

Table 1 Covalent surface immobilization of proteinic affinity ligands [124]

A. Typical activating-and-leaving strategy and reagents:
1. *N*-hydroxysuccinimide esters [NHS]

2. dicyclohexyl carbodiimide [DCC]

Table 1 (continued)

3. water soluble 1-ethyl-3-(3-dimethylaminopropyl) carbodiimide [EDC] with sulfo-NHS

4. N,N'-carbonyldiimidazole [CDI]

5. sulfonyl chlorides: tosyl chloride, mesyl chloride, and tresyl chloride.

6. 2-fluoro-1-methylpyridiniumtoluene-4-sulfonate [FMP]

B. Typical multi-functional coupling strategy and reagents:
1. divinyl sulfone

Table 1 (continued)

2. cyanuric chloride (trichloro-s-triazine)

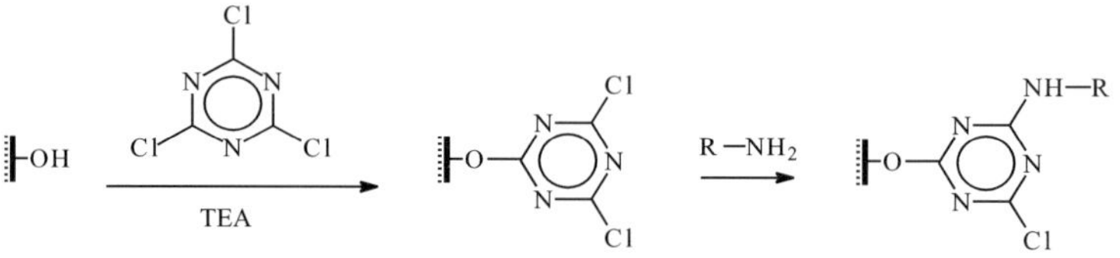

3. diisocyanate

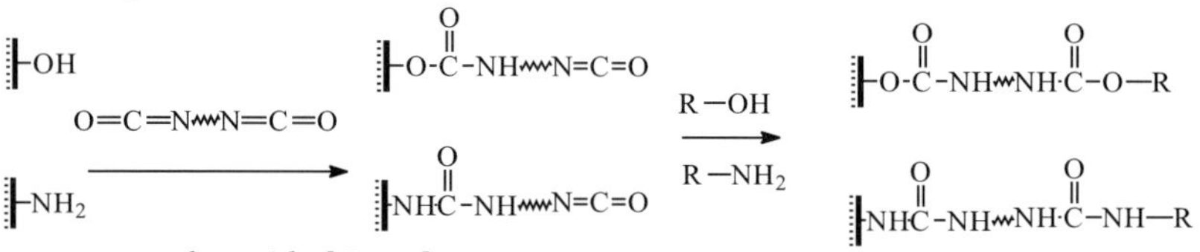

4. cyanogen bromide [CNBr]

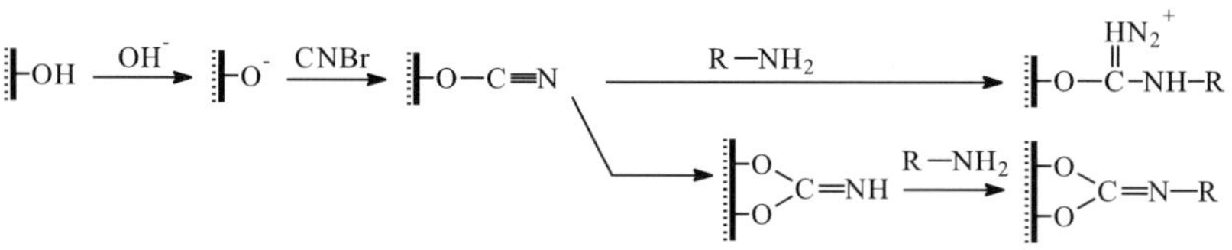

5. azlactone

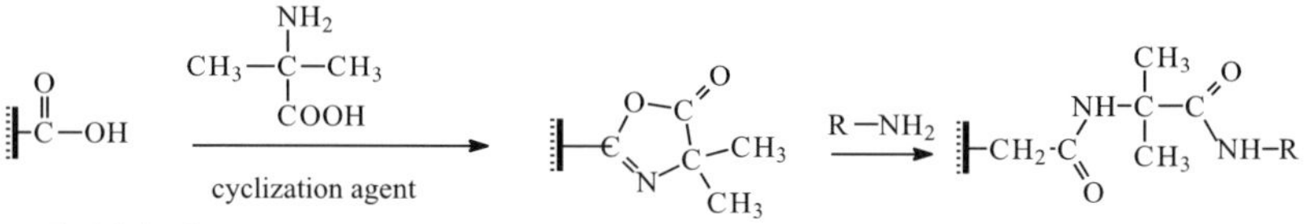

6. dialdehyde

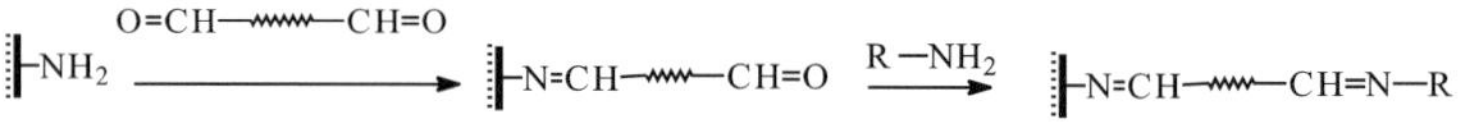

7. glycidol-modification and periodate-activation (aldehyde generation)

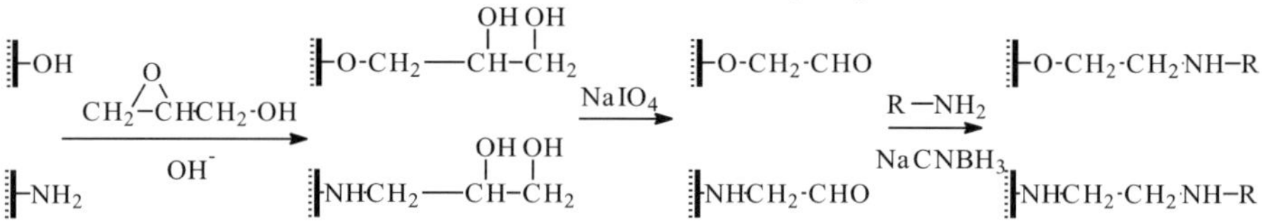

are demonstrated in Table 1B with the specific reagents of divinylsulfone, cyanuric chloride (trichloro-s-triazine), diisocyanate, cyanogen bromide [CNBr], azlactone, dialdehyde, glycidol (epoxy)-modification-and-periodate-activation, and epoxy resin, etc. Typical examples of combined strategies are indicated in Table 1C [113–124]. Some coupling reagents are radio-(photo-) reactive. Typical radio-reactive cross-linkers include photo-sensitive azo, benzophenone, and (meth)acrylate, etc. [121–123].

Table 1 (continued)

C. Examples of combined strategy of activating-and-leaving and multi-functional coupling

1. sulfo-NHS plus maleimide

$$\text{⫶-NH}_2 \xrightarrow{\text{NaO}_3\text{S-[sulfo-NHS-CO-}\sim\sim\text{-CH}_2\text{-maleimide]}}_{\text{R-SH}} \text{⫶-NH-CO-}\sim\sim\text{-CH}_2\text{-[succinimide-S-R]}$$

2. EDC plus bromoacetyl (or iodoacetyl) activation

$$\text{⫶-NH}_2 \xrightarrow[\text{EDC}]{\text{Br(o r I)-CH}_2\text{-COOH}} \text{⫶-NH-}\overset{\text{O}}{\overset{\|}{\text{C}}}\text{-CH}_2\text{-Br(o r I)} \xrightarrow{\text{R-SH}} \text{⫶-NH-}\overset{\text{O}}{\overset{\|}{\text{C}}}\text{-CH}_2\text{-S-R}$$

1.5.2
Spacer Arm

Spacer arms, also called "leashes" or "spacers", function for intermediating substrate materials and affinity ligands. Theoretically, any bi-functional linker compounds can be the candidate of a spacer. However, traditional spacer arms usually consist of linear hydrocarbon-based oligomeric chains ranging from 10 atoms to 2000 Da in length. The spacer-bridged linkage is achieved by fixing one end of the spacer arm onto the matrix surface, while maintaining the other end free and covalently immobilizing a functional ligand as the endgroup. The representative spacer species are hydrocarbon di-amines or oligo-imines, AAs or oligo-peptides (AAs), and aminated epoxides or oligo-ethers [124].

Spacer components themselves are also capable of making some independent contributions, such as surface steric exclusion for non-specific adhesions and modulations of surface hydrophilic/hydrophobic or electrical affinity, etc., however, the major designing role of spacer arms is to support and coordinate the functions of ligand endgroups. Spacer arms help the immobilized ligands to maintain a certain distance from the rigid matrix surface, which avoids the ligands being physically "fastened" and getting denatured by the consequent conformation change. Simultaneously, the use of spacer arms also overcomes the potential steric hindrance and enables specific ligand-receptor corporation. In the structures of ligand-targeted affinity counterparts—whether complementary proteins or other macromolecular cell-surface receptors—the ligand binding sites are usually small moieties that are often buried inside the three-dimensional "pocket-like" macromolecular body. Spacer arms stretch out the pendant ligands as their free endgroups into the "pocket" reaching the receptor binding moieties. To fulfill this purpose, three preconditions are implied: (1) small ligand size; (2) good spacer hydrophilicity; and (3) high spacer-chain flexibility and mobility in the aqueous phase. Concerning the ligand size, it has been discussed in Sect. 1.3.1. Instead of full-sized proteins or other biological macromolecules, oligomeric sequences or compounds are widely recommended and applied. The subsequent challenge is the hydrophilicity of SMA (including ligands and spacers), which dominates the effectiveness and compatibility of the ligands with the (usually aqueous) applying conditions as that in any physiological environments. Obviously, hydrophobic ligands need support at the interface of the water phase; hydrophilic ligands also need to maintain their mobility for receptor-seeking and binding. Hydrophilic and flexible spacers, represented by PEG and polysaccharides, are capable of satisfying these requirements by constructing a fluid-movement-like spacer sub-layer above the material–water interface. Over this sub-layer, the immobilized ligands, as endgroups of spacers, are "floating" freely within a range confined by the spacer length. This model gives the bio-functional ligands ideal activity and optimal mobility.

2
"H-Bond Grafting" Surface Modification for Blood-Contact Polyurethane Biomaterials

2.1
"H-Bond Grafting" Surface Modification for Polyurethane Biomaterials

2.1.1
Model Materials and Surface Modifying Additives [SMA]

The model matrix materials used in this study are thermoplastic polyether urethanes [PEU] (typically using commercially available product Pellethane 2363-80AE, Dow Chemical, Midland, MI, USA). Having excellent tissue compatibility, outstanding mechanical gifts, and lack of potential *staphylococcus*-related infection, PEU has been extensively applied for the fabrication of cardiovascular soft tissue substitutes and other interventional medical devices [125–133]. It is well known that thermoplastic PEU is a linear alternative block-copolymer composed of polyether "soft" blocks and aromatic urethane "hard" blocks. The soft blocks are usually low-molecular weight polymers or oligomers of ether-alkoxyls that intertwine with each other forming a continuous soft phase and provide mechanical flexibility for the overall system. The vast soft phases are connected by PEU hard blocks. In the hard blocks, urethane groups consist of ester-carbonyls and imines that are typical H-bond "chromophores". They are prone to bridge H-bonds between carbonyl-oxygen and imine-protons. Within the bulk of PEU material, abundant H-bonds are founded between the hard blocks localized on different PEU chains and/or different segments of the same chain, by which the linear PEU chains are physically crosslinked into a three-dimensional network. In comparison with the continuous polyether soft phase, the hard block incorporations, namely the physical crosslinking points, aggregate into scattered stiff domains and provide great mechanical modulus for PEU bulk. The microphase-separated structure of PEU, as indicated in Fig. 1, takes responsibility for its elastomeric property and also hints at the SMA-anchoring strategy in this study.

For surface modification, to summarize the conclusions in Sect. 1, successful strategies usually consist of two levels of optimization: optimal SMA design and optimal SMA localization. Combining the discussions in Sects. 1.2.3 and 1.3.2 about functional surface construction, the preference for SMA lo-

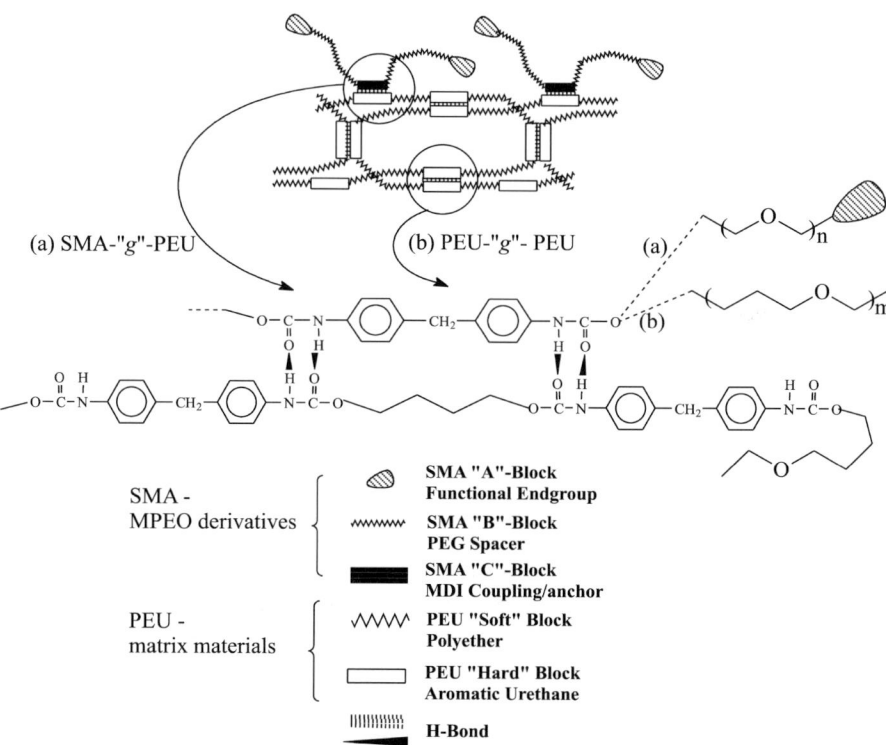

Fig. 1 Schematic illustration of H-bond grafting mechanism. Illustration **a** demonstrates hydrogen-bonded incorporation between MPEO-derived SMA and matrix PEU hard blocks, which mimics, as indicated in Illustration **b**, the similarly hydrogen-bonded self-crosslinking of PEU hard blocks in PEU bulk [76–78, 82]. Reproduced from [173–176]

calization is physical coating (blends). Accordingly, the ideal SMA compound for physical incorporation should be composed of at least three units: bio-functional (end)group, spacer arm, and localizing anchor.

Bio-Functional Endgroup
As concluded in Sects. 1.2.1 and 1.3.1, smaller-sized endgroups with pertinent expertise and appropriate efficiency would always serve the desired functionality better. The details will be depicted in this part of the review.

Spacer Arm
General principles of spacer arms have been thoroughly discussed in Sect. 1.5.2. To fulfill the blood-contact applications in this study, polyethylene glycol [PEG; or polyethylene oxide, PEO] is employed as the only candidate for spacer arms. As a family of well-acknowledged biocompatible compounds, PEG has numerous superior physicochemical, biochemical, and biological properties. (1) PEG is highly water-soluble and also soluble in vast numbers of organic solvents. Among the frequently used solvents, ether and hexane are the only two non-solvents for PEG. This pleasant solubility assures the simplicity and biocompatibility of most PEGylation processes, including the treatments of isolation and purification. (2) PEG is non-toxic and non-immunogenic. The low-molecular-weight PEG-derivatives can be promptly metabolized by the host body. (3) PEG is highly hydrophilic and PEG chains have high mobility and flexibility in compatible phases. If introduced onto biomaterial surfaces, PEG chains are capable of creating large thermodynamic exclusion volumes and simultaneously enhancing the material–medium interface fluidity, as explicated in Sect. 1.5.2, which directly favors the living immobilization of biological ligands by providing physical mobility and hindering non-specific bindings. (4) All these superior properties of PEG can be conveniently rendered to its adjacent compounds or groups via covalent conjugation (PEGylation) without affecting their original characteristics [134–139].

Localizing Anchor
The design of the anchors totally depends on the proposed mechanism of SMA-matrix material incorporation. In this study, 4,4′-methylene diphenyl diisocyanate (MDI) is applied as the localizing anchor that physically bonds to the PEU matrix material by mimicking the physical crosslink between PEU hard blocks. Finally, the "ABCBA"-type SMA, named "MPEO", is constructed as shown in Fig. 1, in which "A" represents bio-functional endgroups, "B" represents the PEG spacer, and "C" represents the MDI anchor that also plays the role of coupling groups for A–B moieties on both sides [76, 77].

2.1.2
SMA "H-Bond Grafting" for Stable Blend and Surface Modification

The general strategy for SMA H-bond grafting is explicated in Fig. 1. The PEU hard blocks are transplanted into the SMA-MPEO structure serving as coupling units ("C"-block in Fig. 1) connecting both PEG spacers ("B"-block in Fig. 1). SMA-MPEO is mixed with PEU chains in a co-solution system. The elastomeric MPEO-PEU blends are achieved by co-solution casting and subsequent solvent evaporation. During the exclusion of solvent, PEU hard blocks gradually crosslink together via H-bond "conjugation", simultaneously microphase separation also occurs by the incorporation of H-bonded hard blocks. Since the "C"-blocks of SMA-MPEO are exactly the copies of PEU hard blocks, what happens to PEU hard blocks also equally happens to the SMA "C"-blocks. Hence, the "C"-blocks, together with the overall compound of MPEO, are physically grafted onto PEU hard block domains via H-bond connection. That's how MDI-based MPEO "C"-blocks function as localizing anchors for SMA immobilization [76, 77].

The H-bond grafting mechanism is certified by infrared spectroscopic evidence and also verified by blending stability evidence. A special MPEO that is endowed with stearic (18-Carbon fatty alkyl) endgroups ("A"-blocks), named "MSPEO", was applied as the model SMA; accordingly, a mono-stearic polyethylene glycol with equivalent PEG chain-length (molecular weight, Mw, $\sim$ 2000 Da), abbreviated "SPEO", was employed as the control. The only difference between MSPEO and SPEO is that SPEO does not possess the MDI anchors.

Infrared Spectroscopic Evidence
Among the regular means of chemical analysis, infrared [IR] spectroscopy specially targets the functional/structural groups of the compounds of interest as information units. Attenuated total reflection Fourier transform infrared spectroscopy [ATR-FT-IR] has the talent of revealing molecular information in the surface layers of materials with an effective detecting depth of 400 nm [140, 141]. Both regular transmission IR and ATR-FT-IR were employed to recruit the information respectively from material bulk and surface layer. To investigate H-bond structures in PEU systems, the IR-targeting unit is the ester-carbonyl groups localized on PEU hard blocks, which, as mentioned above in Sect. 2.1.1, are major H-bond "chromophores". The IR absorbance band of non-H-bonded, or "free" carbonyl groups, namely "– CO – (a)", is centered at $1730\,\mathrm{cm}^{-1}$; while the band of H-bonded (with urethane imine) carbonyl groups, namely "– CO – (b)", is centered at $1700\,\mathrm{cm}^{-1}$ [142, 143]. The quantification follows Lambert–Beer's Law. The ratio of IR absorbance, – CO – (b) to – CO – (a), reflects the relative quantity of H-bonded structures to their non-H-bonded counterparts among all the urethane-containing structures in PEU systems. As shown in Fig. 2, the IR absorbance

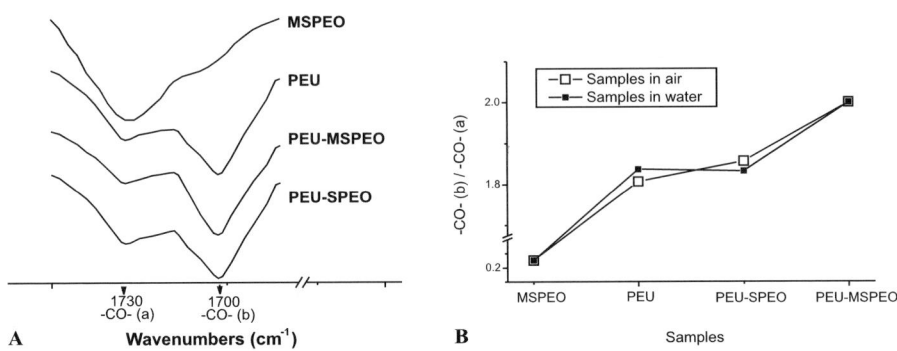

Fig. 2 Identification of H-bond grafting mechanism. **A** ATR-FTIR spectrum highlighting stretch bands of hydrogen-bonded carbonyl groups, – CO – (b), at 1700 cm^{-1}, and free carbonyl groups, – CO – (a), at 1730 cm^{-1}, respectively. **B** ATR-FTIR data curves indicating ratios of – CO – (b) vs – CO – (a) of each sample: SMA-MSPEO, un-modified PEU surfaces, SPEO-modified PEU surfaces, and MSPEO-modified PEU surfaces [77]. Reproduced from [173]

of – CO – (a) and – CO – (b) was respectively measured from the parallel samples of pure MSPEO powders, pure PEU elastomers, PEU-SPEO (5%, w/w) elastomeric blends, and PEU-MSPEO (5%, w/w) elastomeric blends. The quantification is demonstrated with a "– CO – (b)/– CO – (a)" profile that is also exhibited in Fig. 2. The profile indicates that (1) in pure MSPEO material, there are few H-bonded structures formed; (2) in pure PEU elastomer, most of the PEU hard blocks are H-bonded; (3) the addition of SPEO makes little change; however, (4) the addition of MSPEO observably enhances the total quantity of H-bonded structures. Consequently, the conclusion is that, compared to pure PEU, the extra H-bonded structures are generated by the H-bond constructions between PEU hard blocks and MSPEO "C"-blocks, namely, the SMA-MSPEO is physically grafted onto the backbone of PEU matrix materials via H-bond conjugation. The IR spectroscopic evidence is proven to be applicable both in the material bulk and on material surfaces [76, 77].

Blending Stability Evidence

The overall blending stability of SMA in the material bulk and the surface "grafting" stability on material surfaces were examined by leaching tests and evaluated respectively with proton nuclear magnetic resonance spectroscopy [^{1}H-NMR] and quantitative ATR-FT-IR. Firstly, SMA-MSPEO and SPEO with equivalent amounts of PEG components were respectively blended into PEU matrix materials. The initial quantity of PEG was measured and recorded by integrating the PEG-specific ^{1}H-NMR peak areas at $\delta = 3.52$ ppm (– O – C$\underline{H_2}$ – C$\underline{H_2}$ – O –). The integral values were normalized with an internal standard—the area of the ^{1}H-NMR peak at $\delta = 3.36$ ppm ($- O - \underline{CH_2} - CH_2 - CH_2 - \underline{CH_2} - O -$) that represents the (half) content of matrix PEU soft blocks, polytetramethylene glycol [PTMG]. The leaching tests were performed by immersing PEU-MSPEO and PEU-SPEO blends, respectively, (i) in daily refreshed deionized water at 37 °C for 7 days and (ii) in toluene at room temperature for 60 h, followed by incubation in deionized water at 37 °C for 24 h. Under condition (i) at physiological temperature for a week, 85% of SMA persisted in the PEU-MSPEO bulk, while in the PEU-SPEO blends overall only 30% SMA remained. If using a well-known, powerful, PEG-leaching reagent like toluene used for condition (ii), almost all of the SMA-SPEO was washed off from the PEU bulk, while on the contrary in the PEU-MSPEO blends, nearly half of the SMA-MSPEO remained. These evidences confirm the significant contribution for blending stability by the H-bond grafting strategy. A supplementary test modifying condition (i) with an elevated temperature of 70 °C, which is beyond the melting point temperature [Tm] of PEG (66 °C for Mw 2000 Da), indicated an almost entire removal of SMA, no matter whether MSPEO ($< 10\%$ leftover) or SPEO (~ 0 leftover) was involved, from the PEU matrix. This suggests that the H-bond effect is quenched when the polymeric SMA turns into a micro-fluid state [76, 77].

SMA Auto-Makeup Phenomenon
Interestingly, the situation on the material surfaces is quite different to that in the material bulks. Quantitative ATR-FT-IR suggests that, as the PEU-MSPEO blends become sufficiently hydrated (or saturated with toluene) by overnight incubation, the PEG contents on their surfaces always remain constant throughout the whole testing duration, which is independent from the variation of SMA-remainder concentration in the material bulks caused by the different leaching treatments. This phenomenon is not H-bond grafting-related. It implies a surface-oriented migration of SMA from the bulk. The SMA units keep enriching themselves in the surface layer until a containing-saturation is reached. This spontaneous procedure leads to surface occupation of SMA, and thereby fulfills the purpose of surface modification. Moreover, when SMA loss occurs on the material/environment interface, it also tends to maintain the surface content of SMA by the equivalent SMA immigration to reoccupy the vacancies created by the loss. This continuous auto-makeup of SMA further reinforces the stability of surface modification derived by the H-bond grafting strategy [76, 77].

2.1.3
SMA Migration and Surface Enrichment

As described above in Sect. 2.1.2, the localization and immobilization of SMA-M(S)PEO is obtained by the anchoring strategy of H-bond grafting. This mix-in-bulk method only serves for generating a stable blend. The achievement of PEU surface modification relies on spontaneous SMA migration and surface enrichment, which equivalently enables the SMA auto-makeup procedure for SMA's surface reoccupation. The mechanism of SMA spontaneous migration and surface enrichment is explicated as follows.

Preconditions
(i) The typical elastomeric characteristic of the PEU matrix allows the migration of invading environmental small molecules and even SMA macromolecules within PEU bulk; simultaneously it also allows the compliant motion of PEU chain segments towed by SMA via the H-bond connection. (ii) The dissimilarity of mingling compatibility and interface energy between SMA components and the PEU major phase provides both thermodynamic motivation and orientation for SMA migration. (iii) The H-bond grafting conjugation assures the stability of SMA immobilization with PEU matrix.

Driven Force and Orientation
The driven force for SMA-MSPEO migration in PEU bulk originates from the spontaneous micro-phase separation between the SMA-MSPEO components that consist of PEG spacers with stearic endgroups and the major phase of PEU matrix that is constructed by PTMG soft blocks. This micro-phase separation is caused by their unfavorable mutual compatibility. In air, the SMA migration is oriented by the lowest-surface-energy tendency of the stearic endgroups. The air usually behaves as a non-polar hydrophobic phase, hence it has good affinity with non-polar fatty alkyls. The surface occupation of stearic groups helps to minimize the surface energy at the material–air interface. In water, the elastomeric PEU material is slightly swollen by the invasion of water molecules especially in the surface-adjacent area, by which a reducing gradient of water content is formed from the superficial surface to the deep bulk. Given the extremely hydrophilic property of the SMA-PEG spacers, a migration of SMA-MSPEO automatically occurs along with the increasing water-content gradient from the bulk to the PEU–water interface until saturation of the spatial containing capacity is reached in the surface layer. On the water interface, the surface enrichment of hydrophilic PEG-containing SMA also reduces the surface energy, which provides its thermodynamic spontaneity. As mentioned previously in Sect. 2.1.2, the phenomenon of the automatic SMA-surface-reoccupation when loss occurs is also enabled by the same mechanism [144–148].

As discussed above, thermodynamic factors dominate practical SMA behavior, while some kinetic factors also influence the final outcome. For example, kinetic factors such as the molecular size-related steric hindrance and polymeric chain twining are capable of assisting SMA immobilization but retarding SMA migration. Because of the H-bond grafting, whenever SMA migrate, they always move bound with their anchoring PEU segments. Consequently, when the micro-phase separation occurs, the migration of SMA is inevitably encumbered with the connected huge PEU chains, which sterically prevents the SMA migration from aggregating into micelles within the matrix bulk, but just takes a surface-ward orientation [76, 77].

2.1.4
SMA Surface Conformation and Optimization

SMA-based surface architecture fundamentally influences the events and responses of biological substances that approach from the ambient to biomaterial surfaces, thereby directly dominates the overall performance of the surface modifying efforts.

Once SMA is physically enriched on the material surface and exposed to the ambient, the SMA-based surface architecture is mainly determined by the variables of SMA hydrophilicity and the ambient polarity, since these parameters remain the only thermodynamic factors that may alter the interface energy and thus substantially affect the SMA conformation. Accordingly, to explicate the surface topology of the model systems, factorial conditions are manipulated with various hydrophilic/hydrophobic combinations of SMA components versus the ambient. The distinction of ambient polarity has been well acknowledged. Aqueous mediums are undoubtedly of polar circumstances; while the vacuum or air atmosphere is recognized as a non-polar environment with hydrophobic characteristics. For the model SMA, MPEO derivatives, the role of the spacer is always undertaken by PEG chains that possess diverse compatibilities with multiple categories of mediums; in contrast, the candidate selections for SMA endgroups are much more flexible. For example, with the same purpose of pursuing albumin-selective binding on MPEO-decorated PEU surfaces, hydrophilic compounds of Cibacron Blue F3G-A (a triazine dye, "Ciba") and hydrophobic groups of 18-Carbon fatty alkyl (stearyl or "C18") are respectively conjugated to MPEO templates as functional endgroups so as to make use of their specific albumin affinity. The achieved penta-blocked MPEO-derivatives, MSPEO and Ciba-MPEO, are schematically illustrated in Fig. 3. The manipulated SMA surface conformations are also demonstrated in Fig. 3.

As described in Sect. 2.1.3, the tendency towards the lowest surface energy directly dominates all the spontaneous procedures occurring on the SMA–ambient interface, including the construction of SMA surface conformations. On the air interface, MSPEO-decorated PEU surfaces are covered by vast numbers of hydrophobic C18 groups. As revealed by angle-dependent X-ray photoelectron spectroscopy [XPS] (Fig. 4), almost all of these C18 groups are concentrated in the most superficial zone of the surface layer with a depth less than 5 nm. This quantitative information suggests a "standing-up" conformation for the short-chained (Mn 2300 Da) MSPEO on the air interface

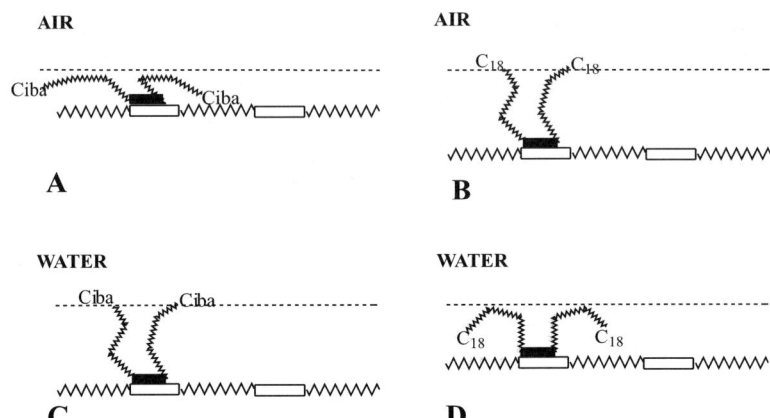

Fig. 3 Schematic illustrations of SMA surface conformations. Cibacron Blue [Ciba]-functionalized MPEO derivatives as SMA on **A** PEU-air and **C** PEU-water interfaces; and stearic group [C_{18}]-functionalized MPEO derivatives as SMA on **B** PEU-air and **D** PEU-water interfaces. Legend for representing symbols applied in schemes is listed as shown in Fig. 1 [77]. Reproduced from [173]

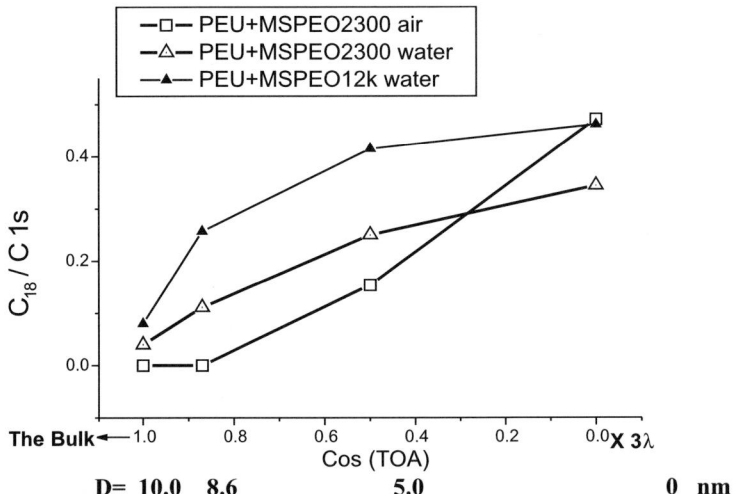

Fig. 4 Distributions of SMA hydrophobic stearic [C_{18}] endgroups vs. hydrophilic PEG spacer arms at various depth of MSPEO-modified PEU surface layer. Ratios of C_{18}/C_{1s} vs. cosine value of XPS takeoff angle [TOA] for SMAs with various-sized PEG spacers (2300 and 12 000 Da) respectively on air and water interfaces, as determined by angle-resolved XPS [76]. Reproduced from [174]

as shown in Fig. 3b. Similar conformations (standing-up shown in Fig. 3c) are also detected from the interface between water and the hydrophilic Ciba-MPEO-modified PEU surfaces. This standing-up conformation reflects a very congenial mixing state between the SMA bodies and the ambient phases. On the contrary, a very compressed conformation of Ciba-MPEO appears on its air interface. As illustrated in Fig. 3a, this "lying-down" pose of Ciba-MPEO is shaped due to the incompatibility between SMA bodies and the ambient air. Interestingly, being different from either the compatible standing-up conformation or the incompatible lying-down conformation, a combined conformation is built up by MSPEO on its water interface. By the influence of ambient water, the hydrophilic PEG spacers tend to stretch out and head to the water phase, simultaneously the hydrophobic C18 endgroups are excluded by the water phase and draw back towards the material bulk. These combined tendencies finally result in a compromised "bending-loop" conformation of MSPEO as demonstrated in Fig. 3d. This conformation is also verified by quantitative XPS analysis and demonstrated in Fig. 4. It indicates that for MSPEO-modified PEU surfaces, when the circumstance is altered from the air to water, the allocation of MSPEO C18 groups observably retreats by expanding the distribution from the original narrow superficial area (depth < 5 nm) back to a much broader zone with the depth of over 10 nanometers [149, 150]. This variation clearly suggests a bending-back action of the C18 endgroups. All these findings are also supported by surface spectroscopic evidences (ART-FT-IR) and surface wettability measurements (contact angle) [76, 77, 94–96, 151–154].

Given the fact that biomaterial surface architecture mainly functions on the water interface, whereas many non-hydrophilic ligands (like C18 groups) are engaged as SMA functional endgroups being introduced onto material surfaces, the compromised surface conformations like the bending-loops (as shown in Fig. 3d) are extensively formed in various SMA-based surface-modifying models. In order to counteract the bending-back tendency of the hydrophobic endgroups and hence assure the exposure of these functional ligands on the water interface, larger-sized hydrophilic spacers (like PEO, Mn 12k Da) are applied. As indicated in Fig. 4, compared to the trials with short-chained spacers (Mn 2300 Da), the attempts with long-chained spacers induce a significant increase of C18 content ranging from 50% to 150% at different depths of the surface layer. Especially at the most superficial level, the C18 enrichment becomes equivalent to that of the MSPEO standing-up conformation. This result suggests that the conformational defect, like that of the bending-loops, can be made up by altering functional fractions in SMA compositions, such as the increase of the hydrophilic PEG moieties in MSPEO. Nevertheless, unilateral enlargement of the spacer size has both pros and cons. The dual dilemmas are that, from the thermodynamics angle, the greater fraction of hydrophilic spacer moieties may reinforce the physical lift for hydrophobic endgroups on the water interface, but it also carries the risk of diluting the functionality by a too heavy amount of spacers; and from the kinetics angle, the longer spacers are capable of offering higher mobility to the hydrophobic endgroups on the water interface, but simultaneously it may also result in a greater kinetic hindrance (repulsion) to the approaching biological substances. On the basis of the rationales of SMA design and function, as long as functional endgroups are involved for a certain purpose, the role of spacer arms is positioned as facilitating machinery mainly responsible for enabling or optimizing the endgroups' functionality and performance. Thereby the criteria for spacer selection, including species characteristic (especially hydrophilicity) and molecular size, should be predominated by the corresponding parameters of the functional endgroups. It is also true for the design of SMA with hydrophilic endgroups, although their inherent standing-up conformation in water exempts the efforts of manipulating thermodynamic lift at the water interface for the endgroups, yet the favorable matching between SMA endgroups and spacer arms is still crucial for yielding and/or maintaining both physical mobility and biological activity for those immobilized functional ligands. Relevant discussions are also listed in Sect. 1.5.2 [76, 77].

2.2
Bio-Functionality of Engineered Polyurethane Surfaces by "H-Bond Grafting"

2.2.1
Coating Applications and In Vivo Tissue Compatibility

To facilitate practical uses, the factorial explorations with blend-in-bulk models are translated into a surface-coating strategy. Despite the diversity of substrate materials, H-bond grafting blends are manipulated into thin layers and coated onto the substrate surfaces, as a result, the surface-modifying functions of the blends are transplanted onto the surface of the object devices without involving substrate factors. Rationales for coating treatments have been discussed in detail in Sect. 1.3.2. Technically, the plating of SMAs is achieved via a popular dip-coating approach. The representative application is performed by coating the mixture of MPEO-derived SMAs together with PEU onto polyester urethane [PEsU] substrates. In comparison with the blend-in-bulk models, PEU no longer plays the role of matrix material, neither endeavors for the overall mechanics, but instead, as a coating aid only takes responsibility for the formation of SMA-containing coating-films as well as their affinity with the substrate materials, namely behaves more like an adhesive. By this means, a greater SMA dosage is allowed in the coating formulation without decaying the bulk modulus. Thereby the coating strategy not only inherits all the surface-modifying talents from the blend-in-bulk prototype, but also yields better modifying efficiency. The coating aids can also be diversified so as to offer various functions. For example, chitosan has been employed as the coating aid mixed with MPEO-derived SMAs and applied on polyurethane intravascular devices with the purpose of lubrication and anti-infection [155–157].

In vivo toxicity of the coating components has been examined in animal models (ICR small rats and SD big rats). For toxicity assessment, MPEO-derived SMAs are directly dosed to the animals by oral feeding and mainline injection; moreover, SMA-coated PEsU devices are transplanted into the animal bodies via subcutaneous imbedding and intravascular implantation [1, 2]. With these settings, the potential toxicological mutations in vivo, including both acute and chronic reactions, are monitored with clinical assays and medical examinations. The tests include general observations, LD_{50}, hematological assays, pathological histology, and serum biochemical examinations. The results consistently indicate normal physiological conditions of the testing animals; thereby the non-toxicity, namely in vivo tissue compatibility of MPEO-derived SMAs is approved. The other aspect of a materials biocompatibility, haemo-compatibility, will be discussed in the following sections [76–79].

2.2.2
Selective Protein Binding

As summarized in Sect. 1.2, physicochemical processes occurring on the blood–biomaterial interface are mostly trigged by the binding actions of forerunner proteins. Particularly for the counteracting manners of thrombogenesis and thromboresistance on foreign surfaces, the binding competition between serum albumin and fibrinogen [Fg] significantly influences the final outcomes. Albumin is the most abundant protein species in serum. Fg, namely coagulation Factor I, also possesses the highest fraction in serum among all the coagulation factors. Surface coverage by albumin is believed and proved to be positive for anti-coagulation; on the contrary, being a core protein for blood clotting, Fg is highly thrombogenic and responsible for the formation of fibrin that directly constructs the insoluble scaffold of thrombus [144–148]. The rationale for thromboresistance at the blood–biomaterial interface is based on a hypothesis that pre-occupancy of albumin can minimize successive fibrinogen attachment. Accordingly, strategies for albumin-selective binding are designed and pursued by introducing albumin-specific affinity to a background that non-discriminatively repulses any protein adsorptions. The dual machineries are facilitated with MPEO-derived SMAs. As stated in Sect. 1.5.2, the PEG spacer arms of the SMAs are capable of repulsing macromolecular adsorptions in a non-specific manner, which proceeds via a combined act of thermodynamic rebound and kinetic detachment. On the other hand, SMA endgroups play the role of albumin-affinity donors. PEG spacers render physical mobility and spatial flexibility to these albumin-specific ligands and enable them to overcome steric hindrances and reach the binding sites of the receptors. The typical models have been described in Sect. 2.1.4, in which C18 and Ciba are respectively conjugated to MPEO templates as functional endgroups. "**C18**". C18 has undoubted affinity with albumin, which has been interpreted with various mechanisms, including the specific "binding sites pockets" model and "Scatchard plot" model, or simply attributed to non-specific hydrophobicity interaction. By any means, the common precondition for C18 to perform as it does lies in, firstly, its free entry to the macromolecular "cage" or "pocket" of albumin structure, on the basis of which the desired hydrophobic interactions can occur inside, since within water ambient the hydrophobic domains of albumin are inevitably embraced inward by the hydrophilic moieties [144–148, 158, 159]. To fulfill serial free actions in the water circumstance and eventually complete the insertion of hydrophobic C18, it totally relies on the upholding effects offered by PEG spacers with their chain hydrophilicity, conformational mobility, and steric flexibility [160–164]. "**Ciba**". Alternatively, Ciba is also enrolled as functional endgroups for MPEO-templates to provide albumin affinity. Ciba is capable of specifically binding human serum albumin [HSA] receptors at the bilirubin site and binding bovine serum albumin [BSA] receptors at the fatty acid site, which are also critically facilitated by PEG spacers [162–165].

ATR-FTIR Assessment

The ATR-FTIR technique has been used for measurement of protein surface binding with relatively large magnitude and indistinctive species [166–169]. The infrared stretch bands for proteins typically vary in three regions: Amide I 1700 ~ 1600 cm^{-1} (covering α-helix conformation at around 1651 cm^{-1} and β-pleated sheet conformation at around 1630 cm^{-1}), Amide II 1600 ~ 1480 cm^{-1}, and Amide III 1350 ~ 1200 cm^{-1} (covering α-helix conformation at around 1300 cm^{-1} and β-pleated sheet conformation at around 1240 cm^{-1}). Given better spectral simplicity (minimal bands overlap) and consistency, Amide II bands centered at 1550 cm^{-1} are usually employed for actual measurements and quantifications. The calculation is conducted with the Tompkin Equation

$$\frac{A}{N} = \frac{n_{21}E_0^2 \varepsilon}{\cos\theta} \int_0^\infty C(z) \exp\left(\frac{-2z}{d_p}\right) dz,$$

where A is the integrated absorbance; N is the number of internal reflections; n_{21} is the ratio of refractive index, n_2/n_1; ε is the integrated molar absorptivity; θ is the incidence angle; $C(z)$ is the concentration as a function of distance z from the interface; and E_0 is the electric field amplitude; d_p is the depth of penetration, $d_p = \lambda/[2\pi(n_1^2 \sin^2\theta - n_2^2)^{1/2}]$, λ is the infrared wavelength. The equation can be simplified as $\Gamma = A_2 C_s d_p/(2A_s)$, where Γ is the density of protein adsorption (μg/cm^2); A_2 and A_s are respectively integrated peak (the IR bands centered at 1550 cm^{-1}) areas for BSA-adsorbed samples and BSA standard solution; C_s is the concentration of standard solution. Schematic illustrations of the testing methodology and the representative outcomes are demonstrated in Fig. 5.

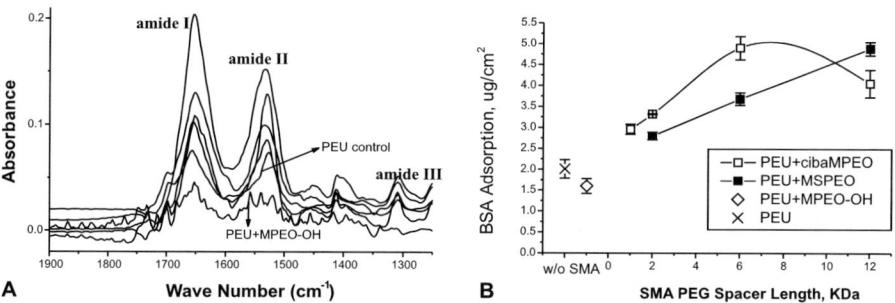

Fig. 5 ATR-FTIR measurement of bovine serum albumin [BSA] adsorption on modified PEU surfaces. **A** ATR-FTIR spectra and **B** quantitative profiles of BSA adsorption on PEU surfaces modified by MPEO-derived SMAs [81]. Reproduced from [177]

The result recalls the desired functioning of MPEO-derived surface modifying machinery. Compared to non-spacer-tethering immobilization, longer spacer arms are capable of offering greater freeness for the conjugated ligands to retain better bio-functional efficacies, until an optimal proportioning between (PEG) spacer length and endgroup species is made. Beyond this proportional threshold, further longer spacers can hardly provide any more positive contributions; on the contrary, the negative effect grows due to the dilution of the endgroup functionalities. The optimal threshold is largely determined by the intrinsic properties of the endgroup ligands, including the physicochemical shape, size, hydrophobicity, as well as relevant biochemical traits. Another important recollection is that the unloaded spacer arms—without functional endgroups, like those on blank MPEO-templates [MPEO-OH], barely function for repulsing any potential adhesion in a non-specific manner so that the BSA binding performance is even weaker than the un-modified PEU surface. Combining these two recollections, the coordination between spacer arms and functional endgroups, as the major feature of MPEO-derived SMAs, is again emphasized and highlighted when pursuing desired bio-functionalities [80, 81].

Radioiodine-Labeling Assessment

The radioiodine (usually radioactive isotope ^{125}I, marked as I*)-labeling technique facilitates protein adsorption assessment with sensitivity and specificity [170], and thereby is employed for evaluations of BSA-vs-Fg binary binding competition as well as reversibility of BSA adsorption. The labeling assessment is capable of identifying and monitoring trace amounts of marked samples from a multi-species system. The labeling methodology is based on radioiodization of target proteins (BSA and/or Fg), which is conducted via a reaction between radioiodine and proteinic tyrosine residues. The radioiodinated proteins (BSA* and/or Fg*) are used for adsorption trials and the binding quantities are measured with a γ-radiation counter.

BSA-vs-Fg Binary Binding Competition

For the purpose of haemo-compatibility, manipulations of protein adsorption aim to maximally resist Fg binding by means of albumin preoccupation. In contrast to the outcomes from the trials with a single protein—no matter BSA or Fg, the setup of BSA-vs-Fg binary binding competition obviously provides a better model to mimic and thus investigate the real reactions occurring at the interface between serum and artificial materials. The experiment is performed with two parallel systems—mixtures of BSA*/Fg and BSA/Fg* with the same concentration fractions that are comparable to the ratio in serum. The results of the competition are presented in Fig. 6a. As compared with the data yielded from untreated PEU materials, the surface decoration by MPEO-derived SMAs simultaneously enhances BSA adsorption and reduces Fg binding in a dramatic manner; in contrast, if solely immobilizing PEG spacers without the presence of functional endgroups (the ligands), as MPEO-OH does, both BSA and Fg bindings are significantly resisted. This result again reveals the SMAs' function of BSA binding efficacy while highlighting the BSA binding specificity against Fg under a competitive condition [80, 81].

BSA Reversible Adsorption

The efficacy of protein adsorption comprises not only the quantity of physical binding but also the maintenance of protein bioactivity. As stated in Sect. 1.5.2, one of the major rationales for granting SMAs with spacer arms is just to maximally avoid denaturing the adsorbed proteins. From the physicochemical angle, a very straightforward means to examine bioactivity of the adsorbed BSA is to test the binding reversibility. Active protein binding is believed to be a dynamic equilibrium between the counteractions of adsorption and desorption, namely, active protein adsorption should be reversible and the adsorbed bioactive proteins should be renewable. The assessment is conducted by exposing sample surfaces to radioiodine-labeled BSA* so-

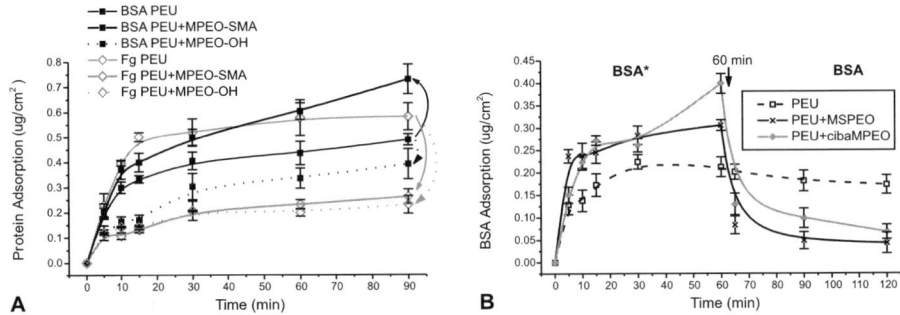

Fig. 6 Competitive binding of bovine serum albumin [BSA] vs. bovine serum fibrinogen [Fg] (**A**), and reversibility of BSA adsorption (**B**), on PEU surfaces modified by MPEO-derived SMAs, as determined by radioiodine labeling [80, 81]. Reproduced from [177, 178]

lution (*Solution A**) until adsorption plateaus (saturation) show up, then trans-subjecting the samples to unlabeled, equally concentrated BSA solution (*Solution B*) for same amount of time. The kinetic curves are presented in Fig. 6b. On the surfaces modified with MPEO-derived SMAs, a renewal of adsorbed BSA is suggested via BSA* desorption in *Solution B*. Since *Solutions A** and *B* have the same BSA concentration, in *Solution B* desorption of *Solution A**-originated BSA* represents a replacement by an equivalent amount of unlabeled BSA from *Solution B*. This type of renewal does not occur on the unmodified PEU surfaces, which implies a denaturation of the adsorbed proteins [80, 81].

2.2.3
Thromboresistance

Blood coagulation consists of extraordinarily complex and delicate processes in which comprehensive species of biological factors are engaged to enable the proceedings through multiple pathways. With a focus on biomaterial-related surface thrombogenesis and thromboresistance, background information has been concisely reviewed in Sect. 1.3. Clotting reactions at the interface of blood and implants are investigated and summarized with a much simplified model as shown in Fig. 7A that outlines a three-stream cascade: (1) plasma pathway referring to fibrin formation, (2) platelet adhe-

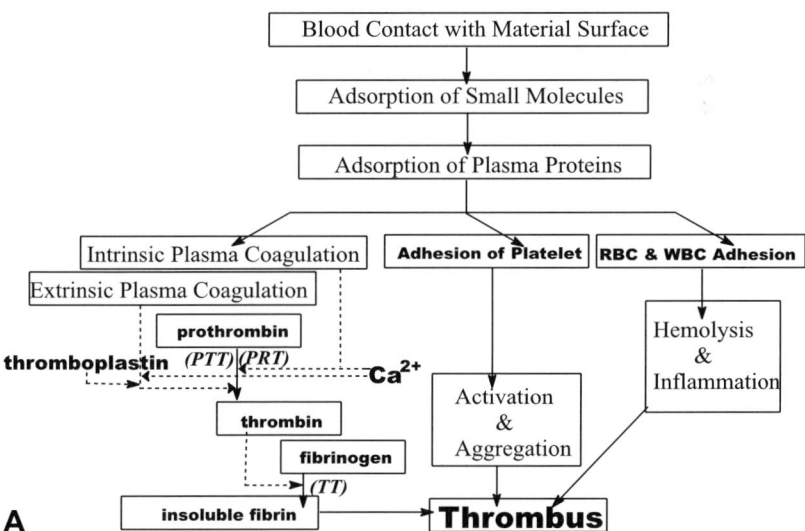

Fig. 7 Blood compatibility of PEU surfaces modified by MPEO-derived SMAs. **A** Simplified cascade model for material-induced blood coagulation highlighting three clotting pathways: plasma fibrin formation, platelet aggregation, and hemolysis-inflammation, respectively characterized by **B–C**

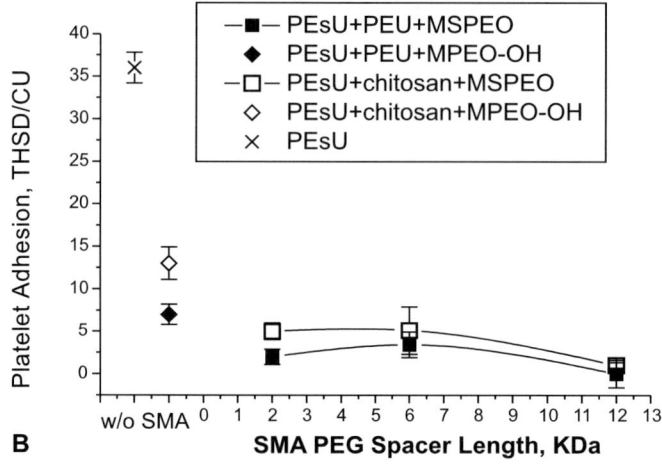

Fig. 7 B Platelet adhesion and aggregation measurement

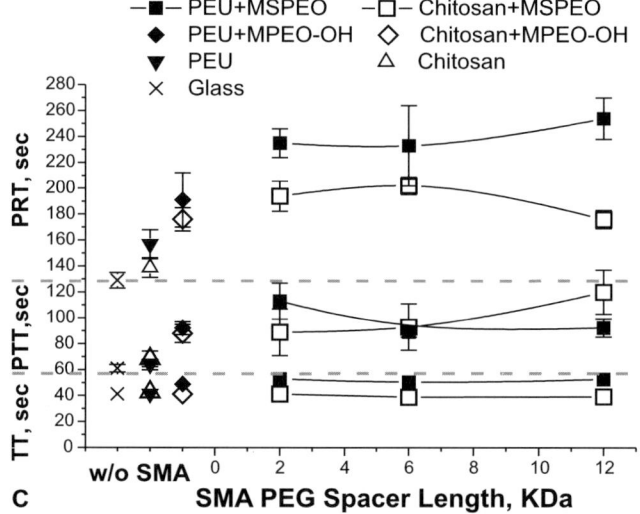

Fig. 7 C Plasma clotting time measurements, including measurements of plasma recalcification time [PRT], plasma thromboplastin-catalyzed clotting time [PTT], and thrombin-catalyzed clotting time [TT]

sion/aggregation, and (3) hemolysis/inflammation. In thrombus, insoluble fibrin constructs the scaffolds, while platelets catalyze the thrombogenic evolution in plasma and cement with other lysed blood cells as the co-fillers [84–86]. Given the purpose of rendering haemo-compatibility to PEsU materials, MPEO-derived SMAs are again employed. The factorial study still mainly focuses on the coordination between PEG spacers and C18 endgroups of

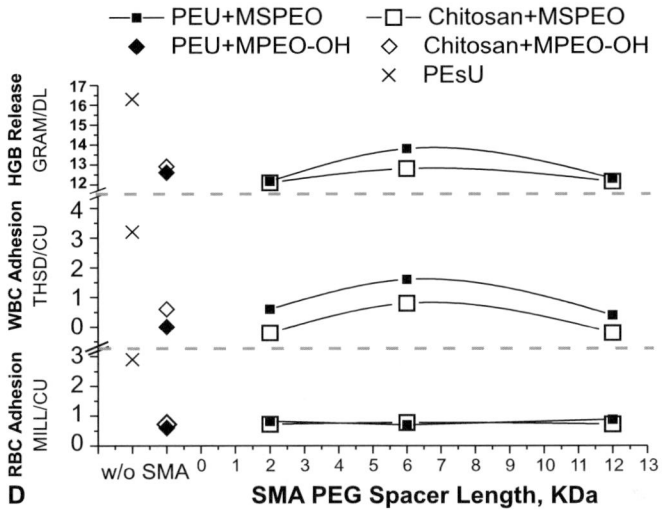

Fig. 7 D Measurements of hemoglobin [HGB] and blood cells including red blood cells [RBC] and white blood cells [WBC] [78]. Reproduced from [179]

the SMAs, and also involves variables of coating aids, respectively PEU and chitosan. Physiological-mimicking blood flows are manipulated with a peristaltic pump system by which various shear rates are simulated and exerted to the modified surfaces for a performance testing under dynamic conditions [171, 172].

Resistance of Platelet Adhesion

Platelets, also called thrombocytes, are plasma membrane-possessed megakaryocyte fragments that play a pivotal role in blood clotting. Generally, damage of native vascular endothelium or invasion of foreign materials tends to activate platelets and thereby lead to platelet adhesion, aggregation, and release. Platelet behavior largely dominates the entire course of the thrombogenic cascade. Initially, the adhesion and aggregation of activated platelets serves for preliminary haemostasis by producing floppy "sealants". Platelet aggregates provide phospholipidic substrates for the activation of bound coagulation factors (typically clotting Factors V and prothrombin), which enables the clotting cascade to proceed via a series of catalyzed surface reactions. The activated platelets are also capable of releasing thrombogenic intracellular substances (typically ADP, etc) to further solidify the clot. Accordingly, in order to pursue haemo-compatible biomaterial surfaces, the SMAs' primary responsibility is to resist platelet adhesion and simultaneously minimize the binding of proteinic clotting factors, so as to hinder the thrombogenic cascade. As indicated in Fig. 7B, MPEO-derived SMAs function for minimizing platelet adhesion on the modified PEsU surfaces and the optimal efficacy is attributed to a proper coordination of SMA spacers with functional endgroups. The surface-modifying efficacy by SMA spacer-endgroup coordination implies a crucial contribution of manipulated albumin preoccupation as described in Sect. 2.2.2 [78, 79].

Retardation of Fibrin Formation
According to the simplified, materials invasion-induced clotting model as shown in Fig. 7A, the plasma coagulation cascade consists of two synchronous and interrelated streams: the intrinsic and extrinsic pathways. Briefly, the intrinsic route is initiated by a series of spontaneous reactions between thrombogenic proteins (in turn, activation of Factors XII, XI, IX, VIII, X, and V) and calcium (Factor IV) in plasma, which finally results in formation of thrombin from prothrombin (Factor II). Thrombin is responsible for catalyzing the transformation of fibrinogen (Fg, Factor I) into soluble fibrin monomers, then polymerized fibrin, finally crosslinked insoluble fibrin (intermediate by activated Factor XIIIa) and thus directly causing plasma coagulation. In comparison with the intrinsic itinerary, the extrinsic way is trigged by "exogenous" (other than plasma components) tissue factors, namely thromboplastin (Factor III) that is usually released from the tissues or cells adjacent to the bleeding injuries and, with the presence of calcium, is capable of accelerating the transformation of thrombin from prothrombin (with the activation of Factors VII and V in turn). Both the intrinsic and extrinsic tracks merge at the spot of thrombin production. Accordingly, the retardation of plasma coagulation on modified PEsU surfaces is characterized step-by-step with various clotting time measurements: (a) *plasma recalcification time* [PRT] for the intrinsic evolution; (b) *plasma thromboplastin-catalyzed clotting time* [PTT] for the extrinsic reactions; and (c) *thrombin-catalyzed clotting time* [TT]. The data of plasma clotting times are presented in Fig. 7C. As compared with un-treated PEsU materials, PRT and PTT on modified surfaces are significantly increased, whereas little alteration is observed from TT measurement. This result suggests that the advantage of surface modification by MPEO-derived SMAs lies in a retardation of thrombin formation, no matter whether for the intrinsic (by PRT) or extrinsic pathway (by PTT). However, if beyond the stage of thrombin formation, namely once thrombin already appears in plasma, the formation of insoluble fibrin and the consequent plasma coagulation would become inevitable. As discussed in the last section, since vast pre-thrombin stage reactions are catalyzed and enabled on the surface of activated platelets, the efficacy of thrombogenic retardation by the SMAs is also crucially attributed to their function of platelet repulsion. The results again demonstrate a functional optimization by the coordination of SMA spacers with their ligand endgroups, which is also consistent with the outcomes from the platelet binding resistance. At the protein level, although the proceeding of Fg-thrombin reaction cannot be deferred by the SMAs, yet the absolute magnitude of proteinic clotting Factor (typically Fg and even prothrombin) adhesion is reduced from

the modified surfaces by the SMA-induced albumin selective binding as described in Sect. 2.2.2, by this means less fibrin formation is still acquired eventually [78, 79].

Anti-Haemolysis/Inflammation
Extrinsic thrombosis frequently complicates with haemolytic and inflammatory reactions especially on the interface of interventional materials. Given the usual risks of these complications, blood cell compatibility of biomaterials is also considered a significant aspect among the overall haemo-compatibility. Haemolysis refers to breakage of red blood cells [RBCs], which causes the release of haemoglobin [HGB] and haemolytic debris that also joins the construction of the thrombus. The biomaterials invasion-related inflammations originate from surface infection of pathogenic microorganisms and the accompanying leucocyte (namely white blood cells [WBCs]) adhesion. Accordingly, the surface-modifying countermeasure against haemolysis and infection is to generally minimize the cell-leveled adhesion. On the basis of the prototype of MPEO-derived SMAs, spatial repulsion and denaturation protection are talents of PEG spacers; while for cell-leveled adsorbing counterparts, C18 endgroups are unable to directly offer any specific binding affinity. Therefore, as demonstrated in Fig. 7D, an unsophisticated resistance is established to both RBC and WBC adhesion on the modified PEsU surfaces [78, 79].

2.2.4
Endothelialization

Tissue-engineered strategies engaged with development of blood-contacting devices are largely represented by surface endothelialization of the employed biomaterials. The rationale is straightforward. An engineered vascular endothelium offers ideal compatibility when interfacing with blood components, hence despite other comprehensive vascular functions and features such as smooth muscle-related vessel mechanics etc., in vivo renewable endothelialization illuminates the ultimate solution for haemo-compatibility of synthetic substitutes. The applied cells for transplantation are usually vascular endothelial cells [ECs]. The seeding-to-confluence of ECs on the inner surface of engineered blood vessels, namely the engineered endothelium, is capable of isolating the blood stream from directly touching the synthetic vascular wall that is mostly vulnerable to clotting. Accordingly, a virtual model of tissue engineered, permanent substitution of blood vessels can be built up with this strategy—the synthetic outer wall is utilized for suture with the connective tissue-based native vascular wall at both ends of the substitute; while the engineered inner endothelium also tends to fuse with the native endothelium from both sides and eventually achieves a complete confluence with the native cell populations [30–36].

Polyurethanes have been widely applied as matrix (wall) materials of synthetic vascular grafts. With the purpose of endothelialization, the main goal of surface modification is to enhance EC affinity for both cell adhesion and proliferation. In this study, MPEO-derived SMAs are enrolled to

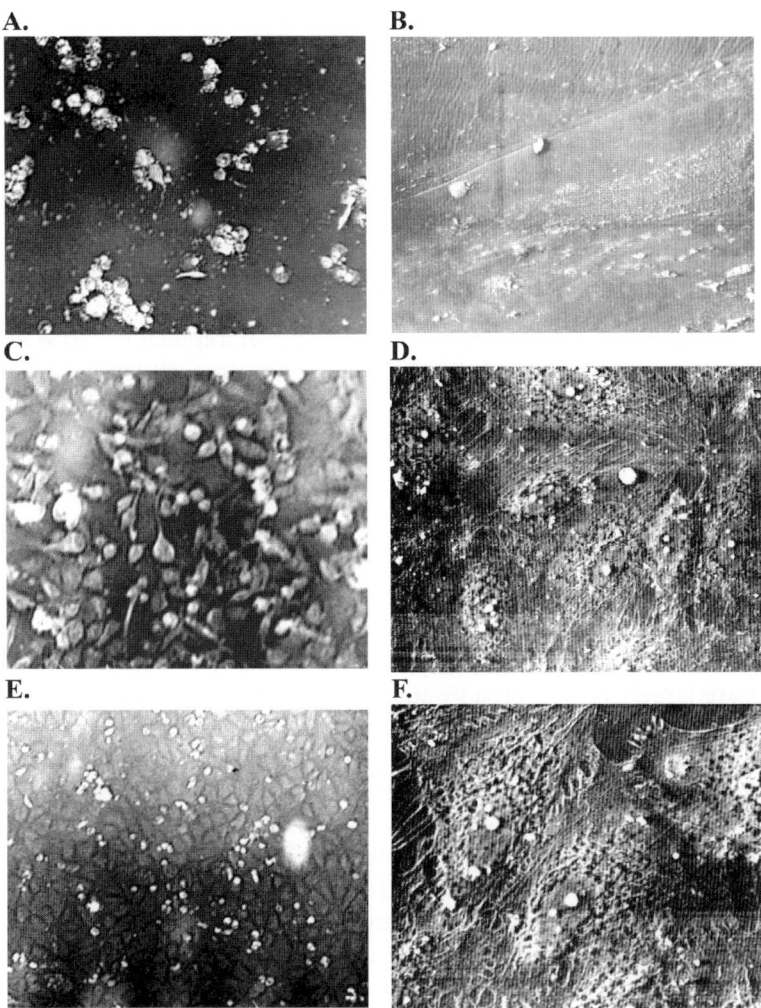

Fig. 8 Endothelialization on PEU surfaces modified by MPEO-derived SMAs. **A** Un-treated PEU surface (control) imaged by optical microscopy [OM]; **B** MPEO-OH (without functional endgroups at the end of PEG spacers) as SMA imaged by scanning electronic microscopy [SEM]; basic amino acid (typically lysine)-functionalized MPEO derivatives as SMA by **C** OM and **D** SEM; arginine-glycin-aspartic acid tri-peptide sequence [RGD]-functionalized MPEO derivatives as SMA by **E** OM and **F** SEM [82, 83]. Reproduced from [180, 181]

optimize interface microenvironments for better cell responses. As discussed in previous sections, a major talent of MPEO-derived SMAs lies in their ability of selectively binding and repulsing the molecular-scaled substances approaching the interface. This ability is attributed to the characteristic of MPEO-derivatives—the intramolecular coordination of functional ligand and spacer-arms. The functional ligand provides desired cell affinity with specificity, which is enabled and associated by the spacers. Proper-sized spacers are capable of unspecifically preventing any undesired adsorption that may harm or dilute the ligand functionality. In the aqueous phase, the spacers are also responsible for physically raising up the ligand and consequently increasing the chance of reaching the receptors on EC surfaces. This coordinating mechanism demonstrates how molecular-leveled manipulation works for cellular events [82, 83].

Technically, the ligands are incorporated to SMAs as endgroups of PEG spacers, followed by a living immobilization onto PEU surfaces. The widely acknowledged cell adhesion ligand, RGD tri-peptide is used as a positive control [31–36, 82, 83, 109, 115]. Mono-amino acids (AAs including acidic aspartic acid [Asp, D], hydrophobic phenylalanine [Phe, F], neutral glycine [Gly, G], and basic lysine [Lys, K] and arginine [Arg, R]) and their various combinations are respectively employed as functional ligands. Primary human umbilical vein endothelial cells [HUVECs] are applied as the experimental cells for transplantation. As exhibited in Fig. 8, successful endothelialization is achieved on the basic AA-MPEO-decorated PEU surfaces. The performance is comparable with the RGD-modified positive controls [82, 83].

3
Conclusions

This review recalls the advances in research on cardiovascular biomaterials, starting from systematic investigations that are summarized centering on an H-bond grafting model that is designed to facilitate surface modification of blood-contact polyurethanes. The development of this model is initiated with originality, and also benefits from ideas and methodologies drawn from comprehensive works from worldwide contributions. The entire project assembles a series of self-contained studies covering aspects from prototype setup through various blood-contact applications. H-bond grafting is an innovative strategy functioning to conjugate SMAs with polyurethane matrix in non-covalent ways such as the blending-coating process. The SMAs are designed as a family of penta-blocked "ABCBA"-type oligomers. The "C"-blocks mimic polyurethane hard blocks in structure, so that in blends of polyurethanes and SMAs, the blenders incorporate each other via hydrogen bonds, which exactly replicates the buildup of physical crosslinking points intrinsically existing in polyurethane elastomers. Namely, the "C"-blocks act as H-bond grafting anchors. The other two components of the SMA body, the "A" and "B" blocks, play the roles of affinity ligands and supporting spacer-arms. The spacers are enrolled with hydrophilic and flexible oligomer chains, typically PEGs. The ligands are immobilized as endgroups of the spacers, the selection of which depends on the required biofunctional specialties. The coordination between SMA endgroups and spacers is responsible for efficient commitment of the endgroup ligands to their corresponding receptors on the target biological counterparts. The modified polyurethane surfaces are used as a platform for assessments of selective protein adsorption, blood-contacting compatibility, and vascular cell hybridization. Accordingly, superior efficacies have been demonstrated by the surface modification on albumin-selective binding, biocompatibility, and engineered endothelialization. Additionally, the technical simplicity of these strategies illuminates a bright perspective for delivery to practical uses.

References

1. Boretos JW, Eden M (1984) Contemporary biomaterials, material and host response, clinical applications, new technology and legal aspects. Noyes Publications, Park Ridge, NJ
2. Helmus NM, Hubbell JA (1993) Cardiovas Path 2:S53
3. Von Recum AF (1986) Handbook of biomaterials evaluation, scientific, technical, and clinical testing of implant materials. Macmillan Press, New York
4. Ratner BD (2000) J Biomater Sci Polym Ed 11:1107
5. Mann BK, West JL (2001) Anat Rec 263:367
6. Haycox CL, Ratner BD (1993) J Biomed Mater Res 27:1181
7. Snyder RW, Helmus MN (1988) Encyclopedia of medical devices and instrumentation, vol 4. Wiley, New York
8. Guidoin R, Marceau D, Rao TJ, King M, Merhi Y, Roy PE, Martin L, Duval M (1987) Biomaterials 8:433
9. Sullivan SJ, Maki T, Borland KM, Mahoney MD, Solomon BA, Muller TE, Monaco AP, Chick WL (1991) Science 252:718
10. Tsai CC, Huo HH, Kulkarni P, Eberhart RC (1990) ASAIO Trans 36:M307
11. Takahara A, Hergenrother RW, Coury AJ, Cooper SL (1992) J Biomed Mater Res 26:801
12. Shimizu T, Kouketsu K, Morishima Y, Goto S, Hasegawa I, Kamiya T, Tamura Y, Kora S (1989) Transfusion 29:292
13. Desai NP, Hubbell JA (1991) J Biomed Mater Res 25:829
14. Williams DF (1981) Biocompatibility of clinical implant materials, vol 2. CRC Press, Boca Raton, Florida
15. Barenberg SA, Brash JL, Narayan R, Redpath AE (1990) Degradable materials: perspectives, issues and opportunities. CRC Press, Boca Raton, Florida
16. Levy RJ, Schoen FJ, Flowers WB, Staelin ST (1991) J Biomed Mater Res 25:905
17. Hoffman AS (1987) Ann NY Acad Sci 516:96
18. Triolo PM, Andrade JD (1983) J Biomed Mater Res 17:149
19. Chenoweth DE (1987) Ann NY Acad Sci 516:306

20. Hubbell JA, Massia SP, Desai NP, Drumheller PD (1991) Biotechnology (NY) 9:568
21. Peppas NA, Langer R (1994) Science 263:1715
22. Langer R, Vacanti JP (1993) Science 260:920
23. Hubbell JA (1995) Biotechnology (NY) 13:565
24. Wojciechowski P, Brash JL (1991) J Biomater Sci Polym Ed 2:203
25. Horbett TA, Lew KR (1994) J Biomater Sci Polym Ed 6:15
26. Chinn JA, Posso SE, Horbett TA, Ratner BD (1992) J Biomed Mater Res 26:757
27. Hynes RO (1992) Cell 69:11
28. Buck CA, Horwitz AF (1987) Annu Rev Cell Biol 3:179
29. Mann BK, West JL (2002) J Biomed Mater Res 60:86
30. Voskerician G, Anderson JM, Ziats NP (2000) J Biomed Mater Res 51:1
31. Mann BK, Tsai AT, Scott-Burden T, West JL (1999) Biomaterials 20:2281
32. Yamada KM (1991) J Biol Chem 266:12809
33. Pierschbacher M, Hayman EG, Ruoslahti E (1983) Proc Natl Acad Sci USA 80:1224
34. Makino M, Okazaki I, Kasai S, Nishi N, Bougaeva M, Weeks BS, Otaka A, Nielsen PK, Yamada Y, Nomizu M (2002) Exp Cell Res 277:95
35. Nomizu M, Weeks BS, Weston CA, Kim WH, Kleinman HK, Yamada Y (1995) FEBS Lett 365:227
36. Varki A (1994) Proc Natl Acad Sci USA 91:7390
37. Lasky LA (1992) Science 258:964
38. Llanos GR, Sefton MV (1993) J Biomater Sci Polym Ed 4:381
39. Amiji M, Park K (1993) J Biomater Sci Polym Ed 4:217
40. Fujimoto K, Tadokoro H, Ueda Y, Ikada Y (1993) Biomaterials 14:442
41. Uchida E, Uyama Y, Ikada Y (1994) Langmuir 10:481
42. Martins CL, Wang DA, Ji J, Feng LX, Barbosa MA (2003) Biomaterials 24:2067
43. Martins CL, Wang DA, Ji J, Feng LX, Barbosa MA (2003) J Biomater Sci Polym Edn 14:439
44. Freij-Larsson C, Nylander T, Jannasch P, Wesslen B (1996) Biomaterials 17:2199
45. Gombotz WR, Wang GH, Horbett TA, Hoffman AS (1991) J Biomed Mater Res 25:1547
46. Merrill EW (1993) J Biomater Sci Polym Edn 5:1
47. Osterberg E, Bergstrom K, Holmberg K, Schuman TP, Riggs JA, Burns NL, Van Alstine JM, Harris JM (1995) J Biomed Mater Res 29:741
48. Burns NL, Vanalstine JM, Harris JM (1995) Langmuir 11:2768
49. Lopez GP, Ratner BD, Tidwell CD, Haycox CL, Rapoza RJ, Horbett TA (1992) J Biomed Mater Res 26:415
50. Nojiri C, Okano T, Koyanagi H, Nakahama S, Park KD, Kim SW (1992) J Biomater Sci Polym Edn 4:75
51. Ishihara K, Fukumoto K, Iwasaki Y, Nakabayashi N (1999) Biomaterials 20:1545
52. Furuzono T, Ishihara K, Nakabayashi N, Tamada Y (2000) Biomaterials 21:327
53. Terlingen JG, Brenneisen LM, Super HT, Pijpers AP, Hoffman AS, Feijen J (1993) J Biomater Sci Polym Edn 4:165
54. Tseng YC, McPherson T, Yuan CS, Park K (1995) Biomaterials 16:963
55. Kuhl TL, Leckband DE, Lasic DD, Israelachvili JN (1994) Biophys J 66:1479
56. Ji J, Feng LX, Shen JC (2003) Langmuir 19:2643
57. Prime KL, Whitesides GM (1993) J Am Chem Soc 115:10714
58. Lopez GP, Albers MW, Schreiber SL, Carroll R, Peralta E, Whitesides GM (1993) J Am Chem Soc 115:5877
59. Uchida M, Tanizaki T, Oda T, Kajiyama T (1991) Macromolecules 24:3238
60. Lasic DD (1994) Angew Chem Int Ed Engl 33:1685

61. Dimilla PA, Folkers JP, Biebuyck HA, Harter R, Lopez GP, Whitesides GM (1994) J Am Chem Soc 116:2225
62. Ferguson GS, Chaudhury MK, Biebuyck HA, Whitesides GM (1993) Macromolecules 26:5870
63. O'Shea GM, Sun AM (1986) Diabetes 35:943
64. Sawhney AS, Hubbell JA (1992) Biomaterials 13:863
65. Ishihara K, Fukumoto K, Iwasaki Y, Nakabayashi N (1999) Biomaterials 20:1545
66. Ishihara K, Fukumoto K, Iwasaki Y, Nakabayashi N (1999) Biomaterials 20:1553
67. Schaub RD, Kameneva MV, Borovetz HS, Wagner WR (2000) J Biomed Mater Res 49:460
68. Suggs LJ, West JL, Mikos AG (1999) Biomaterials 20:683
69. Silver JH, Myers CW, Lim F, Cooper SL (1994) Biomaterials 15:695
70. Brunstedt MR, Ziats NP, Robertson SP, Hiltner A, Anderson JM, Lodoen GA, Payet CR (1993) J Biomed Mater Res 27:367
71. Grasel TG, Castner DG, Ratner BD, Cooper SL (1990) J Biomed Mater Res 24:605
72. Yoon SC, Ratner BD, Ivan B, Kennedy JP (1994) Macromolecules 27:1548
73. Tingey KG, Andrade JD (1991) Langmuir 7:2471
74. Stauffer SR, Peppas NA (1992) Polymer 33:3932
75. Chaikof EL, Merrill EW, Callow AD, Connolly RJ, Verdon SL, Ramberg K (1992) J Biomed Mater Res 26:1163
76. Wang DA, Ji J, Feng LX (2000) Macromol Chem Phys 201:1574
77. Wang DA, Ji J, Feng LX (2000) Macromolecules 33:8472
78. Wang DA, Ji J, Gao CY, Yu GH, Feng LX (2001) Biomaterials 22:1549
79. Wang DA, Ji J, Sun YH, Yu GH, Feng LX (2001) J Biomed Mater Res 58:372
80. Wang DA, Ji J, Feng LX (2001) J Biomater Sci Polym Edn 12:1123
81. Wang DA, Chen BL, Ji J, Feng LX (2002) Bioconjugate Chem 13:792
82. Wang DA, Ji J, Sun YH, Shen JC, Feng LX, Elisseeff JH (2002) Biomacromolecules 3:1286
83. Wang DA, Feng LX, Ji J, Sun YH, Zheng XX, Elisseeff JH (2003) J Biomed Mater Res 65A:498
84. Plow EF, Herren T, Redlitz A, Miles LA, Hoover-Plow JL (1995) FASEB J 9:939
85. Remes A, Williams DF (1992) Biomaterials 13:731
86. Esmon CT (1995) FASEB J 9:946
87. March J (1985) Advanced organic chemistry, 3rd edn. Wiley, New York
88. Anderson AB, Tran TH, Hamilton MJ, Chudzik SJ, Hastings BP, Melchior MJ, Hergenrother RW (1996) AJNR Am J Neuroradiol 17:859
89. Hardhammar PA, Van Beusekom HM, Emanuelsson HU, Hofma SH, Albertsson PA, Verdouw PD, Boersma E, Serruys PW, Van der Giessen WJ (1996) Circulation 93:423
90. Serruys PW, Emanuelsson H, van der Giessen W, Lunn AC, Kiemeney F, Macaya C, Rutsch W, Heyndrickx G, Suryapranata H, Legrand V, Goy JJ, Materne P, Bonnier H, Morice MC, Fajadet J, Belardi J, Colombo A, Garcia E, Ruygrok P, de Jaegere P, Morel MA (1996) Circulation 93:412
91. Sauvage LR, Berger KE, Wood SJ, Yates SG 2nd, Smith JC, Mansfield PB (1974) Arch Surg 109:698
92. Robinson KA, Li J, Mathison M, Redkar A, Cui J, Chronos NA, Matheny RG, Badylak SF (2005) Circulation 112(9 Suppl):I135
93. Cui J, Li J, Mathison M, Tondato F, Mulkey SP, Micko C, Chronos NA, Robinson KA (2005) Cardiovasc Revasc Med 6:113
94. Ueda T, Ishihara K, Nakabayashi N (1995) J Biomed Mater Res 29:381

95. Ishihara K, Aragaki R, Ueda T, Watenabe A, Nakabayashi N (1990) J Biomed Mater Res 24:1069
96. Grasel TG, Pierce JA, Cooper SL (1987) J Biomed Mater Res 21:815
97. Gamble JR, Harlan JM, Klebanoff SJ, Vadas MA (1985) Proc Natl Acad Sci USA 82:8667
98. Cameron BL, Tsuchida H, Connall TP, Nagae T, Furukawa K, Wilson SE (1993) J Cardiovasc Surg (Torino) 34:281
99. Clowes AW, Kohler T (1991) J Vasc Surg 13:734
100. Jarrell BE, Williams SK (1991) J Vasc Surg 13:733
101. Graham LM, Fox PL (1991) J Vasc Surg 13:742
102. Schmedlen RH, Elbjeirami WM, Gobin AS, West JL (2003) Clin Plast Surg 30:507
103. Elbjeirami WM, Yonter EO, Starcher BC, West JL (2003) J Biomed Mater Res A66: 513
104. Zilla P, Deutsch M, Meinhart J (1999) Semin Vasc Surg 12:52
105. Magometschnigg H, Kadletz M, Vodrazka M, Dock W, Grimm M, Grabenwoger M, Minar E, Staudacher M, Fenzl G, Wolner E (1992) J Vasc Surg 15:527
106. Gray JL, Kang SS, Zenni GC, Kim DU, Kim PI, Burgess WH, Drohan W, Winkles JA, Haudenschild CC, Greisler HP (1994) J Surg 57:596
107. Kang SS, Gosselin C, Ren D, Greisler HP (1995) Surgery 118:280
108. Gosselin C, Ren D, Ellinger J, Greisler HP (1995) Am J Surg 170:126
109. Fittkau MH, Zilla P, Bezuidenhout D, Lutolf MP, Human P, Hubbell JA, Davies N (2005) Biomaterials 26:167
110. Seliktar D, Zisch AH, Lutolf MP, Wrana JL, Hubbell JA (2004) J Biomed Mater Res A68:704
111. Zisch AH, Lutolf MP, Ehrbar M, Raeber GP, Rizzi SC, Davies N, Schmokel H, Bezuidenhout D, Djonov V, Zilla P, Hubbell JA (2003) FASEB J 17:2260
112. Greisler HP, Dennis JW, Endean ED, Ellinger J, Buttle KF, Kim DU (1988) Circulation 78(3Pt2):I6
113. Castillo EJ, Koenig JL, Anderson JM, Lo J (1985) Biomaterials 6:338
114. Drumheller SD, Hubbell JA (1995) Surface immobilization of adhesion ligands for investigations of cell substrate interactions. In: Bronzino JD (ed) The biomedical engineering handbook. CRC Press, Boca Raton, Fl, p 1584
115. Beer JH, Springer KT, Coller BS (1992) Blood 79:117
116. Coller BS, Beer JH, Scudder LE, Steinberg MH (1989) Blood 74:182
117. Kobayashi H, Hyon S H, Ikada Y (1991) J Biomed Mater Res 25:1481
118. Kobayashi H, Ikada Y (1991) Biomaterials 12:747
119. Liu SQ, Ito Y, Imanishi Y (1993) J Biomed Mater Res 27:909
120. Werb Z, Tremble PM, Behrendtsen O, Crowley E, Damsky CH (1989) J Cell Biol 109:877
121. Noh I, Goodman SL, Hubbell JA (1998) J Biomater Sci Polym Edn 9:407
122. Yan M, Cai SX, Wybourne MN, Keana JFW (1993) J Am Chem Soc 115:814
123. Suzuki Y, kusakabe M, Iwaki M (1991) Nucl Instrum Meth B 59:1300
124. Hermanson GT, Mallia AK, Smith PK (1992) Immobilized affinity ligand techniques. Academic Press, London
125. Stankus JJ, Guan J, Fujimoto K, Wagner WR (2006) Biomaterials 27:735
126. Stankus JJ, Guan J, Wagner WR (2004) J Biomed Mater Res A70:603
127. Guan J, Wagner WR (2005) Biomacromolecules 6:2833
128. Guan J, Fujimoto KL, Sacks MS, Wagner WR (2005) Biomaterials 26:3961
129. Guan J, Sacks MS, Beckman EJ, Wagner WR (2004) Biomaterials 25:85
130. Guan J, Sacks MS, Beckman EJ, Wagner WR (2002) J Biomed Mater Res 61:493

131. Sanders JE, Lamont SE, Karchin A, Golledge SL, Ratner BD (2005) Biomaterials 26:813
132. Lelah MD, Cooper SL (1986) Polyurethanes in medicine. CRC Press, Boca Raton, Florida
133. Ikada Y (1984) Adv Polym 57:103
134. Massia SP, Hubbell JA (1992) Cytotechnology 10:189
135. Brunstedt MR, Ziats NP, Schubert M, Stack S, Rose-Caprara V, Hiltner PA, Anderson JM (1993) J Biomed Mater Res 27:499
136. Rao SB, Sharma CP (1997) J Biomed Mater Res 34:21
137. Tarsi R, Muzzarelli RA, Guzman CA, Pruzzo C (1997) J Dent Res 76:665
138. Han DK, Park K, Ryu G, Kim UY, Min BG, Kim YH (1996) J Biomed Mater Res 30:23
139. Aldenhoff YB, Koole LH (1995) J Biomed Mater Res 29:917
140. Ohta K, Iwamoto R (1985) Appl Spectrosc 39:418
141. Ohta K, Iwamoto R (1985) Anal Chem 57:2491
142. Coleman MM, Lee KH, Skrovanek DJ, Painter PC (1986) Macromolecules 19:2149
143. Harthcock MA (1989) Polymer 30:1234
144. Ji J, Feng LX, Qiu YX, Yu XJ, Barbosa MA (2000) J Colloid Interface Sci 224:255
145. Ji J, Feng LX, Qiu YX, Yu XJ (2000) Polymer 41:3713
146. Pitt WG, Cooper SL (1988) J Biomed Mater Res 22:359
147. Pitt WG, Grasel TG, Cooper SL (1988) Biomaterials 9:36
148. Pitt WG, Cooper SL (1986) Biomaterials 7:340
149. Jannasch P (1998) Macromolecules 31:1341
150. Jannasch P (2000) Macromolecules 33:8604
151. Desai NP, Hubbell JA (1991) Biomaterials 12:144
152. Andrade JD (1985) Surface and interfacial aspects of biomedical polymers, vol 1, surface chemistry and physics. Plenum Press, New York
153. Gregonis DE, Hsu R, Buerger DE, Smith LM, Andrade JD (1982) Macromolecular solutions. Pergamon Press, London
154. Marco C, Fatou JG, Gomez MA, Tanaka H, Tonelli AE (1990) Macromolecules 23:2183
155. Kojima K, Okamoto Y, Kojima K, Miyatake K, Fujise H, Shigemasa Y, Minami S (2004) J Vet Med Sci 66:1595
156. Morimoto M, Saimoto H, Usui H, Okamoto Y, Minami S, Shigemasa Y (2001) Biomacromolecules 2:1133
157. Minami S, Okamoto Y, Hamada K, Fukumoto Y, Shigemasa Y (1999) EXS 87:265
158. Spector AA (1975) J Lipid Res 16:165
159. Goodman DS (1958) J Am Chem Soc 80:3892
160. McCormick RM, Karger BL (1980) Anal Chem 52:2249
161. Park KD, Kim YS, Han DK, Kim YH, Lee EH, Suh H, Choi KS (1998) Biomaterials 19:851
162. Keogh JR, Wolf MF, Overend ME, Tang L, Eaton JW (1996) Biomaterials 17:1987
163. Keogh JR, Eaton JW (1994) J Lab Clin Med 124:537
164. Keogh JR, Velander FF, Eaton JW (1992) J Biomed Mater Res 26:441
165. Leatherbarrow RJ, Dean PD (1980) Biochem J 189:27
166. Chittur KK (1998) Biomaterials 19:357
167. Lenk TJ, Ratner BD, Gendreau RM, Chittur KK (1989) J Biomed Mater Res 23:549
168. Jeon JS, Sperline RP, Raghavan S (1992) Appl Spectrosc 46:1644
169. Jeon JS, Raghavan S, Sperline RP (1994) Colloids Surf A 92:255
170. Brash JL, Davidson VJ (1979) Thromb Res 9:249
171. Podias A, Groth T, Missirlis Y (1994) J Biomater Sci Polym Edn 6:399

172. Patel JD, Iwasaki Y, Ishihara K, Anderson JM (2005) J Biomed Mater Res A73:359
173. Wang DA et al. (2000) Macomolecules 33:8476–8477
174. Wang DA et al. (2000) Macromol Chem Phys 201:1575
175. Wang DA et al. (2001) Biomaterials 22:1553
176. Wang DA et al. (2002) Biomacromolecules 3:1287
177. Wang DA et al. (2002) Bioconjugate Chem 13:798
178. Wang DA et al. (2001) J Biomater Sci Polym Ed 12:1136–1137
179. Wang DA et al. (2001) Biomaterials 22:1555
180. Wang DA (2002) Biomacromolecules 3:1291, 1294
181. Wang DA et al. (2003) J Biomed Mater Res 65A:507–508

Editor: Karel Dušek

Author Index Volumes 201–209

Author Index Volumes 1–100 see Volume 100
Author Index Volumes 101–200 see Volume 200

Alekseeva, T., see Lipatov, Y. S.: Vol. 208
Anwander, R. see Fischbach, A.: Vol. 204, pp. 155–290.
Ayres, L. see Löwik D. W. P. M.: Vol. 202, pp. 19–52.

Binder, W. H. and *Zirbs, R.*: Supramolecular Polymers and Networks with Hydrogen Bonds in the Main- and Side-Chain. Vol. 207, pp. 1–78
Bouteiller, L.: Assembly via Hydrogen Bonds of Low Molar Mass Compounds into Supramolecular Polymers. Vol. 207, pp. 79–112
Boutevin, B., David, G. and *Boyer, C.*: Telechelic Oligomers and Macromonomers by Radical Techniques. Vol. 206, pp. 31–135
Boyer, C., see Boutevin B: Vol. 206, pp. 31–135
ten Brinke, G., Ruokolainen, J. and *Ikkala, O.*: Supramolecular Materials Based On Hydrogen-Bonded Polymers. Vol. 207, pp. 113–177

Csetneki, I., see Filipcsei G: Vol. 206, pp. 137–189

David, G., see Boutevin B: Vol. 206, pp. 31–135
Deming T. J.: Polypeptide and Polypeptide Hybrid Copolymer Synthesis via NCA Polymerization. Vol. 202, pp. 1–18.
Dong Liu, X., Yamada, M., Matsunaga, M. and *Nishi, N.*: Functional Materials Derived from DNA. Vol. 209, pp. 149–178
Donnio, B. and *Guillon, D.*: Liquid Crystalline Dendrimers and Polypedes. Vol. 201, pp. 45–156.

Elisseeff, J. H. see Varghese, S.: Vol. 203, pp. 95–144.
Esker, A. R., Kim, C. and *Yu, H.*: Polymer Monolayer Dynamics. Vol. 209, pp. 59–110

Ferguson, J. S., see Gong B: Vol. 206, pp. 1–29
Filipcsei, G., Csetneki, I., Szilágyi, A. and *Zrínyi, M.*: Magnetic Field-Responsive Smart Polymer Composites. Vol. 206, pp. 137–189
Fischbach, A. and *Anwander, R.*: Rare-Earth Metals and Aluminum Getting Close in Ziegler-type Organometallics. Vol. 204, pp. 155–290.
Fischbach, C. and *Mooney, D. J.*: Polymeric Systems for Bioinspired Delivery of Angiogenic Molecules. Vol. 203, pp. 191–222.
Freier T.: Biopolyesters in Tissue Engineering Applications. Vol. 203, pp. 1–62.
Friebe, L., Nuyken, O. and *Obrecht, W.*: Neodymium Based Ziegler/Natta Catalysts and their Application in Diene Polymerization. Vol. 204, pp. 1–154.

García A. J.: Interfaces to Control Cell-Biomaterial Adhesive Interactions. Vol. 203, pp. 171–190.
Gong, B., Sanford, AR. and *Ferguson, JS.*: Enforced Folding of Unnatural Oligomers: Creating Hollow Helices with Nanosized Pores. Vol. 206, pp. 1–29
Guillon, D. see Donnio, B.: Vol. 201, pp. 45–156.

Harada, A., Hashidzume, A. and *Takashima, Y.*: Cyclodextrin-Based Supramolecular Polymers. Vol. 201, pp. 1–44.
Hashidzume, A. see Harada, A.: Vol. 201, pp. 1–44.
Häußler, M. and *Tang, B. Z.*: Functional Hyperbranched Macromolecules Constructed from Acetylenic Triple-Bond Building Blocks. Vol. 209, pp. 1–58
Heinze, T., Liebert, T., Heublein, B. and *Hornig, S.*: Functional Polymers Based on Dextran. Vol. 205, pp. 199–291.
Heßler, N. see Klemm, D.: Vol. 205, pp. 57–104.
Van Hest J. C. M. see Löwik D. W. P. M.: Vol. 202, pp. 19–52.
Heublein, B. see Heinze, T.: Vol. 205, pp. 199–291.
Hornig, S. see Heinze, T.: Vol. 205, pp. 199–291.
Hornung, M. see Klemm, D.: Vol. 205, pp. 57–104.

Ikkala, O., see ten Brinke, G.: Vol. 207, pp. 113–177

Jaeger, W. see Kudaibergenov, S.: Vol. 201, pp. 157–224.
Janowski, B. see Pielichowski, K.: Vol. 201, pp. 225–296.

Kataoka, K. see Osada, K.: Vol. 202, pp. 113–154.
Kim, C., see Esker, A. R.: Vol. 209, pp. 59–110
Klemm, D., Schumann, D., Kramer, F., Heßler, N., Hornung, M., Schmauder H.-P. and *Marsch, S.*: Nanocelluloses as Innovative Polymers in Research and Application. Vol. 205, pp. 57–104.
Klok H.-A. and *Lecommandoux, S.*: Solid-State Structure, Organization and Properties of Peptide—Synthetic Hybrid Block Copolymers. Vol. 202, pp. 75–112.
Kosma, P. see Potthast, A.: Vol. 205, pp. 151–198.
Kosma, P. see Rosenau, T.: Vol. 205, pp. 105–149.
Kramer, F. see Klemm, D.: Vol. 205, pp. 57–104.
Kudaibergenov, S., Jaeger, W. and *Laschewsky, A.*: Polymeric Betaines: Synthesis, Characterization, and Application. Vol. 201, pp. 157–224.

Laschewsky, A. see Kudaibergenov, S.: Vol. 201, pp. 157–224.
Lecommandoux, S. see Klok H.-A.: Vol. 202, pp. 75–112.
Li, S., see Li W: Vol. 206, pp. 191–210
Li, W. and *Li, S.*: Molecular Imprinting: A Versatile Tool for Separation, Sensors and Catalysis. Vol. 206, pp. 191–210
Liebert, T. see Heinze, T.: Vol. 205, pp. 199–291.
Lipatov, Y. S. and *Alekseeva, T.*: Phase-Separated Interpenetrating Polymer Networks. Vol. 208
Löwik, D. W. P. M., Ayres, L., Smeenk, J. M., Van Hest J. C. M.: Synthesis of Bio-Inspired Hybrid Polymers Using Peptide Synthesis and Protein Engineering. Vol. 202, pp. 19–52.
Lucas, P. and *Robin, J.-J.*: Silicone-Based Polymer Blends: An Overview of the Materials and Processes. Vol. 209, pp. 111–147

Marsch, S. see Klemm, D.: Vol. 205, pp. 57–104.
Matsunaga, M., see Dong Liu, X.: Vol. 209, pp. 149–178

Mooney, D. J. see Fischbach, C.: Vol. 203, pp. 191–222.

Nishi, N., see Dong Liu, X.: Vol. 209, pp. 149–178
Nishio Y.: Material Functionalization of Cellulose and Related Polysaccharides via Diverse Microcompositions. Vol. 205, pp. 1–55.
Njuguna, J. see Pielichowski, K.: Vol. 201, pp. 225–296.
Nuyken, O. see Friebe, L.: Vol. 204, pp. 1–154.

Obrecht, W. see Friebe, L.: Vol. 204, pp. 1–154.
Osada, K. and *Kataoka, K.*: Drug and Gene Delivery Based on Supramolecular Assembly of PEG-Polypeptide Hybrid Block Copolymers. Vol. 202, pp. 113–154.

Pielichowski, J. see Pielichowski, K.: Vol. 201, pp. 225–296.
Pielichowski, K., Njuguna, J., Janowski, B. and *Pielichowski, J.*: Polyhedral Oligomeric Silsesquioxanes (POSS)-Containing Nanohybrid Polymers. Vol. 201, pp. 225–296.
Pompe, T. see Werner, C.: Vol. 203, pp. 63–94.
Potthast, A., Rosenau, T. and *Kosma, P.*: Analysis of Oxidized Functionalities in Cellulose. Vol. 205, pp. 151–198.
Potthast, A. see Rosenau, T.: Vol. 205, pp. 105–149.

Robin, J.-J., see Lucas, P.: Vol. 209, pp. 111–147
Rosenau, T., Potthast, A. and *Kosma, P.*: Trapping of Reactive Intermediates to Study Reaction Mechanisms in Cellulose Chemistry. Vol. 205, pp. 105–149.
Rosenau, T. see Potthast, A.: Vol. 205, pp. 151–198.
Rotello, V. M., see Xu, H.: Vol. 207, pp. 179–198
Ruokolainen, J., see ten Brinke, G.: Vol. 207, pp. 113–177

Salchert, K. see Werner, C.: Vol. 203, pp. 63–94.
Sanford, A. R., see Gong B: Vol. 206, pp. 1–29
Schlaad H.: Solution Properties of Polypeptide-based Copolymers. Vol. 202, pp. 53–74.
Schmauder H.-P. see Klemm, D.: Vol. 205, pp. 57–104.
Schumann, D. see Klemm, D.: Vol. 205, pp. 57–104.
Smeenk, J. M. see Löwik D. W. P. M.: Vol. 202, pp. 19–52.
Srivastava, S., see Xu, H.: Vol. 207, pp. 179–198
Szilágyi, A., see Filipcsei G: Vol. 206, pp. 137–189

Takashima, Y. see Harada, A.: Vol. 201, pp. 1–44.
Tang, B. Z., see Häußler, M.: Vol. 209, pp. 1–58

Varghese, S. and *Elisseeff, J. H.*: Hydrogels for Musculoskeletal Tissue Engineering. Vol. 203, pp. 95–144.

Wang, D.-A.: Engineering Blood-Contact Biomaterials by "H-Bond Grafting" Surface Modification. Vol. 209, pp. 179–227
Werner, C., Pompe, T. and *Salchert, K.*: Modulating Extracellular Matrix at Interfaces of Polymeric Materials. Vol. 203, pp. 63–94.

Xu, H., Srivastava, S., and *Rotello, V. M.*: Nanocomposites Based on Hydrogen Bonds. Vol. 207, pp. 179–198

Yamada, M., see Dong Liu, X.: Vol. 209, pp. 149–178
Yu, H., see Esker, A. R.: Vol. 209, pp. 59–110

Zhang, S. see Zhao, X.: Vol. 203, pp. 145–170.
Zhao, X. and *Zhang, S.*: Self-Assembling Nanopeptides Become a New Type of Biomaterial. Vol. 203, pp. 145–170.
Zirbs, R., see Binder, W. H.: Vol. 207, pp. 1–78
Zrínyi, M., see Filipcsei G: Vol. 206, pp. 137–189

Subject Index

AAO template 48
Acetylene-metal reactions 48
Acetylenes, addition reactions 12
–, cobalt carbonyls 49
–, repetitive coupling 5
Acetylenic triple-bond building blocks 1
Activating-and-leaving reagents 196
ADP 190
Adsorption, amphiphilic 187
Affinity ligands, proteinic 196
Ag nanowire 155
Aging-resistance 185
Alkyne homopolycyclotrimerization 20
Alkyne, homo-/cross-couplings 5
Amines, aromatic 165
–, secondary 4
Anionic polymer, helical 152
Anti-adhesion coating 191
Anti-coagulating/thrombolytic drugs 190
Anti-thrombine III (AT-III) 193
Anti-thrombogenic coating 190
Aptamers 163, 171
Arginine–vasopressin 171
Aroylacetylene, cyclotrimerization 33
Aroyldiynes 34
Aryl ethynyl ketone, polycyclotrimerization 35
Aryl halides, repetitive coupling 5
Aryldiynes, (co)polycyclotrimerization 27
AuNPs, DNA hybrids 161

Benzo[a]pyrene 166
Benzophenone, photoreactivity 44
Bicomponent blends, silicone/polymer 114
Bioactive coating 191
Biocompatibility 182
Bioconjugation 195

Biomaterials 179
–, cardiovascular applications 182
Biopolymers 171
Bioresponses 185
Biphenyl 166
Bis(aroylacetylene)s 33
Blends, silicone/polymer 114
Block copolymers 98
Blood compatibility/haemo-compatibility 179, 182
Blood rheology 193
Blood-contact polyurethane biomaterials 200
Blood–implant interaction 189
Bottle-brush copolymers 121

Capillary waves 59
–, dynamics 65
N,N'-Carbonyldiimidazole (CDI) 196
Carcinogenicity, intercalation into DNA 165
Cardiovascular biomedical materials 182
Cardiovascular medical applications, polymeric biomaterials 182
Cardiovascular-functional polymers 183
CdS nanorods 158
Cell adhesion, biological recognition 186
Cellulose acetate butyrate (CAB) 134
Chromophores 200
Click polymerization 15
Coal tar 165
Coalescence prevention 118
Coatings 191
–, anti-thrombogenic 190
–, bioactive 191
Collagen 155
– DNA complex 156
– fibrin 195
– PET 195

Colloids, DNA hybids 161
Compatibilization 111
Copolymer addition, preformed 119
Copolymer formation, graft 123
–, in-situ 122
Copolymers 92
–, alternating 92
–, block 98
–, branched, in-situ 126
Copolyynes 11, 38
Cortin 190
Creep-resistance 185

Daunorubicin 171
Degradation resistance 185
Dibenzofuran 166
Dibenzo-p-dioxin 166
β,β-Dibromo-4-ethynylstyrene 6
Dicumyl peroxide 121, 125
Dicyclohexyl carbodiimide [DCC] 196
Diels–Alder polycycloaddition 14
Diethynylbenzene,
  copolycyclotrimerization 30
Diiodophenylacetylene 5
Dilational elastic modulus 59
Dilational loss modulus 59
3,4-Dimethyl-3,4-diphenylhexane 137
Dioxins 167
Dispersion equation 59
Diynes, transition metal-catalyzed
  polycyclotrimerization 19
DNA binder 149
– cationic lipids 156
– collagen 155
– device 163
– immobilized bead column 167
– inorganic matter 153
– junctions/crossover tiles 159
– ligands, toxic/carcinogenic 165
– mediated molecular structures 161
– molecular combing 154
– motifs, complex 160
– nanoarchitectures,
  "junction"/"crossover" 158
DNA nanomechanical devices 164
DNA nanostructure 158
DNA polymerization, rolling-circle 162
DNA repair 173
DNA strand pairing 158
DNA walker 165

DNA–alginic acid hybrid 169
DNA–chitosan 172
DNA–gelatin 172
DNA–lipid cast film 156
DNA–membrane templates, self-assembled 157
DNA–metal ion complex 153
DNA–metal nanoparticles, oligofunctional 161
DNA–polyacrylamide hydrogel beads 169
Double helix 149
– binding 165

Elastic modulus, dilational 59
Electrophoretic stretching 154
Endocrine disruptors 165, 168
Endothelial cell reoccupation, host 194
Endothelial cell seeding, allogenic 194
Endothelial-derived relaxing factor 193
Endothelialization 179, 219
–, surface-engineered 192
Endothelium, functions 192
Ethidium bromide 166
1-Ethyl-3-(3-dimethylaminopropyl)
  carbodiimide (EDC) 196

Fibrinogen 187
Fibroblast growth factors 195
2-Fluoro-1-methylpyridinium
  toluene-4-sulfonate (FMP) 196
Functional materials 1

Glaser–Hay oxidative coupling route 11
Groove binding/intercalation 165

H-bond grafting mechanism 201
– surface modification 200
HDPE-g-VTES compatibilizer 121
Heart valves, biocompatibility 183
Heparin 190
Heparin-like polysaccharides 193
Homopolymers 80
Homopolyynes 11, 38
Honeycomb morphologies 47
Hydrodynamic stretching 154
Hydrogen bond 179
13-Hydroxy stearic diene acid 193
N-Hydroxy succinimide esters (NHS) 196
Hyperbranched polymers 1
Hyperplasia 193

Subject Index

Implants, artificial, engineered endothelialization 193
Incision anastomosis 184
Inert materials 189
Intercalation 149
IPN membranes, rubber-hydrogel 131
IPNs (interpenetrating polymer networks) 111, 128
IPNs, latex 129, 135
IPNs, semi- 136
IPNs, sequential 129, 130
IPNs, simultaneous 129, 134
IPNs, synthesis 130

Latex IPNs/semi-IPNs 135
LDPE/PDMS, branched 126
LDS/PCS IPNs 133
Leukotrienc 190
Lipid–DNA complexes 157
Lipids, cationic 156

MA-g-EPDM 127
Matrix bulk blending 188
Matrix materials 200
Methyldiethynylsilane 13
4,4′-Methylene diphenyl diisocyanate (MDI) 202
Micrococcal nuclease 166
Molecular combing 154
Molecule device 149
Monolayers 59
–, binary, side chain length effect 88
–, dynamics 104
–, static properties 61

Nanostructure 149
Nanotubes 48
Nanowires 153
NLO chromophores 42

Octamethylcyclotetrasiloxane 117
Octaphenylcyclotetrasiloxane 114

PA6/PDMS 123
PAHs 165
PCBs 165
PCDDs 165
PCDFs 165
PCR 150

PDMS/cellulose acetate butyrate (CAB) IPN 134
PDMS/polyacrylate IPNs, damping zone 132
PDMS-g-polyacrylamide/sulfonated-EPDM 128
PEGylation 202
PEO/PDMS 121
PET 193
Photonic properties 48
Photoreactivity 44
PMPS/PMMA IPN 133
Polyaddition 1
Polyarylenes, hyperbranched 26
Polybutadiene–polydimethylsiloxane 117
Polycarbosilane 13
Polycoupling 1, 5
–, copper-catalyzed 9
–, palladium-catalyzed 5
Polycycloaddition 4, 14
–, 1,3-dipolar (click polymerization) 15
Polycycloaddition, Diels–Alder 14
Polycyclotrimerization 1, 19
Polydimethylsiloxane, polyamide 114
Polyester urethane (PEsU) 210
Polyether urethanes (PEU), thermoplastic 200
Polyethers 80
Polyethylene glycol (PEG) 187, 202
Polyethylene oxide (PEO) 202
Polyhydrosilylation 4, 12
–, palladium-catalyzed 13
–, rhodium-/palladium-catalyzed 13
Polyisoprene (PIP)/PDMS 117
Polymer blends, copolymers, compatibilization 118
–, silicone-based 111
Polymers, hyperbranched 1, 20, 26
Polysaccharides 187
Polysiloxane, poly(ethylene-co-methacrylate) 125
– crosslinking systems 129
Polysiloxane–polycarbonate 114
Polytriazoles, hyperbranched 15
Polyurethane 179
– biomaterials, blood-contact 200
– surfaces, bio-functionality, H-bond grafting 210
Polyvinylpyridine (PVP) 118

Poly(alkylenephenylene)s, hyperbranched 20
Poly(amic ester)-based photoresists 44
Poly(aroylarylene)s, hyperbranched 33
Poly(t-butyl methacrylate) 80
Poly(butylene oxide) 80
Poly(ethylene oxide) 80
Poly(ethylene terephthalate) (PET) 193
Poly(ethylene-co-methacrylate), polysiloxane 125
Poly(ethylene-co-propylene)/PA-6,6 124
Poly(HEMA)/PDMS 131
Poly(2-hydroxyethyl methacrylate) 131
Poly(methyl acrylate) 80
Poly(methyl methacrylate) 80
Poly(silylenedivinylene)s 14
Poly(tetraflouroethylene), expanded (ePTFE) 193
Poly(tetrahydrofuran) 80
Poly(tetramethylene oxide) 115
Poly(urethanes), thermoplastic (TPU) 115
Poly(vinyl acetate) 80
Poly(vinylphenol)/poly(vinyl acetate) 123
PPO/PDMS IPN 135
Prostacyclin 193
Prostaglandin (6-keto-PGE$_1$) 193
Protein adsorption 179, 185
Protein arrays, DNA templated 163
Protein binding, biomaterial surface construction 187
–, selective 211
Protein immobilization 195
Proteinic affinity ligands 196
–, surface immobilization 195
Proteins, DNA hybrids 161
PS/PDMS 115
PSf membrane, DNA 172
PU/PDMS 115
–, semi-IPN 139

Re-endothelialization promoting coating 191
Relaxing factor, endothelial-derived 193
Rolling-circle DNA polymerization 162

Scaling exponent 59
Semi-IPNs (pseudo-IPNs) 129

Serotonin 190
Silicone/polymer blend 111
–, functionalized 117
–, unmodified 114
Silicone-based polymer blends 111
Silicone crosslinking 129
Silicone latex IPNs 135
Silicone rubber/hydrogels 131
Silicone thermoplastic vulcanizate (STPV) 136
Silole-containing polymer 40
Silyldiynes, homopolycyclotrimerization 29
SMA (surface-modifying additives) 200
–, H-bond grafting 203
–, migration/surface enrichment 205
–, surface conformation/optimization 207
SMA-MPEO 203
Spacer arm 199
Stretching, electrophoretic/hydrodynamic 154
Sulfonyl chlorides 196
Surface bioconjugation 195
Surface covalent grafting 187
Surface immobilization, covalent 196
Surface light scattering 59, 75
Surface modification 179
Surface optimization 185
Surface self-assembly 187
Surface thrombogenesis/thromboresistance 189
Surface-modifying additives (SMA) 200
Surfactants 118

Tantallacyclopentadienes 22
Tantalum halides 4
Tetraphenylcyclopentadienones, ethynyl-substituted 14
Th1–Th2 balance 174
Therapy, engineered 182
Thermoplastic elastomers (TPE) 136
Thermoplastic vulcanizates (TPV) 136
Thrombogenesis/thromboresistance 189
Thrombolytic functions 193
Thrombolytic promoters 193
Thrombomodulin (TM) 193
Thromboresistance 189, 193, 215

Subject Index

Thrombospondin 187
Thromboxane 190
Tissue compatibility 182
–, polyurethane 210
Tissue factor pathway inhibitor (TFPI) 193
Tissue plasminogen activator (t-PA) 193
1,3,5-Triethynylbenzene 14
Triynes, (co)polycoupling 10
Two-plane orientation complex 18

Vascular permeability 192
Vascular smooth muscle relaxation factors 193
Vinyl polymers 80
Vinyltriethoxysilane (VTES) 121
Vitamin B_{12} 131
Vitronectin 187
Vulcanizates, thermoplastic (TPV) 136

Willebrand factors 187

Printing: Krips bv, Meppel
Binding: Stürtz, Würzburg

RETURN TO: CHEMISTRY LIBRARY
100 Hildebrand Hall • 510-642-3753

LOAN PERIOD 1	2	3
4	2 HOUR	

~~ALL BOOKS MAY BE RECALLED AFTER 7 DAYS.~~
~~Renewals may be requested by phone, or on GLADIS, type **inv**
followed by your patron ID number~~

DUE AS STAMPED BELOW.

DEC 20

FORM NO. DD 10
1,000 7-07

UNIVERSITY OF CALIFORNIA, BERKELEY
Berkeley, California 94720–6000